CompTIA
Project+®
Study Guide
Exam PK0-004
Second Edition

Kim Heldman

SYBEX®
A Wiley Brand

Senior Acquisitions Editor: Kenyon Brown
Development Editor: James A. Compton
Technical Editor: Vanina Mangano
Production Editor: Dassi Zeidel
Copy Editor: Kim Wimpsett
Editorial Manager: Mary Beth Wakefield
Production Manager: Kathleen Wisor
Executive Editor: Jim Minatel
Book Designers: Judy Fung and Bill Gibson
Proofreader: Kathy Pope, Word One New York
Indexer: Ted Laux
Project Coordinator, Cover: Brent Savage
Cover Designer: Wiley
Cover Image: ©Jeremy Woodhouse/Getty Images, Inc.

Copyright © 2017 by John Wiley & Sons, Inc., Indianapolis, Indiana

Published simultaneously in Canada

ISBN: 978-1-119-28052-1

ISBN: 978-1-119-28054-5 (ebk.)

ISBN: 978-1-119-28053-8 (ebk.)

Manufactured in the United States of America

For general information on our other products and services or to obtain technical support, please contact our Customer Care Department within the U.S. at (877) 762-2974, outside the U.S. at (317) 572-3993 or fax (317) 572-4002.

Wiley publishes in a variety of print and electronic formats and by print-on-demand. Some material included with standard print versions of this book may not be included in e-books or in print-on-demand. If this book refers to media such as a CD or DVD that is not included in the version you purchased, you may download this material at http://booksupport.wiley.com. For more information about Wiley products, visit www.wiley.com.

Library of Congress Control Number: 2016960612

10 9 8 7 6 5 4 3 2 1

To Kate and Juliette, project managers in the making.

Acknowledgments

Thank you for buying the second edition of *CompTIA Project+ Study Guide Exam PK0-004* to help you study and prepare for the CompTIA Project+ exam. I believe this book is a good introduction to the in-depth world of project management and certification and will open up many opportunities for you.

I would like to thank all the great team members at Wiley who were part of this project: Kenyon Brown, senior acquisitions editor; Jim Compton, development editor; Dassi Zeidel, production editor; and all those behind the scenes who helped make this book a success. They are terrific to work with, as always, and I appreciate their keen eyes and insightful ideas and suggestions.

Special thanks go to Vanina Mangano for her work as technical editor. I appreciate her diligence and great suggestions that helped make the content stronger.

And a thank-you, as always, goes to my family for their understanding of my crazy schedule. Kate and Juliette, you're the best!

About the Author

Kim Heldman, MBA, PMP® is the CIO for the Regional Transportation District in Denver, Colorado. Kim directs IT resource planning, budgeting, project prioritization, and strategic and tactical planning. She directs and oversees IT design and development, enterprise resource planning systems, IT infrastructure, application development, cybersecurity, IT program management office, intelligent transportation systems, and data center operations.

Kim oversees the IT portfolio of projects ranging from small in scope and budget to multimillion-dollar, multiyear projects. She has more than 25 years of experience in information technology project management. Kim has served in a senior leadership role for more than 18 years and is regarded as a strategic visionary with an innate ability to collaborate with diverse groups and organizations, instill hope, improve morale, and lead her teams in achieving goals they never thought possible.

Kim is also the author of *PMP® Project Management Professional Exam Study Guide, 8th Edition*; *Project Management JumpStart 3rd Edition*; and *Project Manager's Spotlight on Risk Management*. She is the coauthor of *PMP® Project Management Professional Exam Deluxe Study Guide, 2nd Edition*, and the *PMP® Project Management Professional Exam Review Guide, 3rd Edition*. Kim has also published several articles and is currently working on a leadership book.

Kim continues to write on project management best practices and leadership topics, and she speaks frequently at conferences and events. You can contact Kim at Kim.Heldman@gmail.com. She personally answers all her email.

Contents at a Glance

Contents

CompTIA.

Becoming a CompTIA Certified IT Professional is Easy

It's also the best way to reach greater professional opportunities and rewards.

Why Get CompTIA Certified?

Growing Demand

Project management is one of the business world's most in demand skill sets. Many employees need understand project management skills. While not every individual is a project leader, all individuals need to understand the fundamental concepts of project management.

Higher Salaries

Professionals with certifications on their resume command better jobs, earn higher salaries and have more doors open to new multi-industry opportunities.

Verified Strengths

91% of hiring managers indicate CompTIA certifications are valuable in validating expertise, making certification the best way to demonstrate your competency and knowledge to employers.**

Universal Skills

Professionals who would like to embark on a project management career path possess skills that can be used in virtually any industry – from information technology to consumer goods to business services. This career mobility ensures that project managers can readily find work in any industry.

 Learn 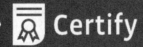 Certify 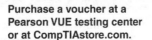 Work

Learn more about what the exam covers by reviewing the following:

- Exam objectives for key study points.

- Sample questions for a general overview of what to expect on the exam and examples of question format.

- Visit online forums, like LinkedIn, to see what other IT professionals say about CompTIA exams.

Purchase a voucher at a Pearson VUE testing center or at CompTIAstore.com.

- Register for your exam at a Pearson VUE testing center:

- Visit pearsonvue.com/CompTIA to find the closest testing center to you.

- Schedule the exam online. You will be required to enter your voucher number or provide payment information at registration.

- Take your certification exam.

Congratulations on your CompTIA certification!

- Make sure to add your certification to your resume.

- Check out the CompTIA Certification Roadmap to plan your next career move.

Learn more: **Certification.CompTIA.org/projectplus**

* Source: CompTIA 9th Annual Information Security Trends study: 500 U.S. IT and Business Executives Responsible for Security
** Source: CompTIA Employer Perceptions of IT Training and Certification

Introduction

Have you ever wondered how the pyramids were built? Or the Eiffel Tower? How did someone have the organizational skills to put all those people together and create such magnificent structures? Coming forward to recent times—how is Microsoft capable of putting together literally *millions* of lines of code for its latest operating system? The answer to all of these is project management.

The CompTIA Project+ exam will test your knowledge of the concepts and processes involved in project management. There are several project management methodologies you can follow, each with their own processes and procedures, but at the foundation of each one are sound project management principles and techniques. CompTIA Project+ is vendor neutral. It acknowledges other methodologies such as those published by the Project Management Institute® and PRINCE2 but doesn't follow them precisely.

In this edition of *CompTIA Project+ Study Guide*, you'll find plenty of discussion of project management concepts such as defining the requirements, creating the project charter document, creating the scope document, planning the project, assessing and managing risk, and closing out the project. You'll also find exam questions in categories such as team building and personnel management, quality management, status reporting, and more, and these are discussed as well.

The Project+ certification used to be centered on information technology (IT) projects and was called the IT Project+ exam. Many projects involve IT in one way or another, so you'll find references in this book to IT-related projects. I should also mention that my job-related experience is in IT, and many of my examples are drawn from real-life situations.

Where should you go beyond taking your Project+ test? If you find you're interested in all things project management, you should enroll in a good university-level class that takes you through the heavier stages of project management. This book and this test only touch the surface of project management techniques. You'll find there is much more to learn and that it's possible to make a career out of managing projects.

 Don't just study the questions and answers in this book; the questions on the actual exam will be different from the practice ones included in the book and the online test bank at www.wiley.com/go/sybextestprep. The exam is designed to test your knowledge of a concept or objective, so use this book to learn the objective behind the question.

What Is the Project+ Certification?

CompTIA's mission is to create tests and certifications that aren't company-specific. For example, you can take a server test that deals with the elements of servers and server operation but doesn't ask you specifics about Dell, HP, or IBM equipment. CompTIA got its start with what is now an industry standard, the A+ exam. This is a test designed to quiz you on your understanding of the workings of a PC and its associated connection to a network. But there are other tests as well: Network+, Linux+, and others.

Why Become Project+ Certified?

Certification in project management has increasingly become a requirement for those interested in full-time careers in this field. It improves your credibility with stakeholders and customers. Becoming certified demonstrates your intent to learn the processes associated with project management and provides you with opportunities for positions and advancements that may not otherwise be possible.

Here are some reasons to consider the Project+ test and this study guide:

Demonstrates Proof of Professional Achievement   Certification demonstrates to current and potential employers that you are knowledgeable and well-grounded in project management practices and have taken the initiative to prove your knowledge in this area.

Increases Your Marketability   If you take a moment to browse job postings for project management positions, you'll often find that certification is either highly desirable or required. The CompTIA Project+ certification will help you stand out from other candidates and demonstrate that you have the skills and knowledge to fulfill the duties required of a project manager.

Provides Opportunity for Advancement   You may find that your Project+ certification is just what you need to get that next step up the ladder. People who study and pass certification tests prove, if nothing else, that they have the tenacity to get through a difficult subject and to prove their understanding by testing on the subject.

Provides a Prerequisite for Advanced Project Management Training   If you're considering a project management career, the Project+ exam is a great way to start. Studying for this exam gives you the background on what project management is really all about, not simply what one company or organization thinks it's about. After passing the exam, you should consider obtaining the Certified Associate Project Management (CAPM®) or Project Management Professional (PMP®) certification through the Project Management Institute. This study guide follows the principles and processes outlined by PMI® and is a great introduction to its certifications.

Raises Customer Confidence Because It Raises Your Confidence   Customers who know you're certified in project management and who hear you speak and act with confidence are more confident in the company you represent. If you're able, for example, to identify and describe the four categories of risks to prepare for on a project, your customer gains confidence in you.

How to Become Project+ Certified

First, study the topics and processes outlined in this book, and make certain to answer all the end-of-chapter questions and take the extra bonus exams that are provided as part of the online test bank at www.wiley.com/go/sybextestprep.

Then go to the CompTIA website (www.comptia.org) to find the list of testing sites where the exam is currently conducted.

Prices and testing centers are subject to change at any time. Please visit CompTIA's website for the most up-to-date information: www.comptia.org.

You'll need a driver's license and one other form of ID when you arrive at the testing center. No calculators, computers, cell phones, or other electronic devices are allowed in the testing area. You're allowed 90 minutes to take the exam, and there are 90 multiple-choice questions. There are no prerequisites for this exam. You'll be notified of your grade as soon as you finish the test.

Who Should Buy This Book?

You should buy this book if you're interested in project management and want to learn more about the topic. The Project+ exam is an ideal way to introduce yourself to project management concepts and techniques, and this book will immerse you in the basics of those techniques.

If you've never taken a certification test before, you'll find that the Project+ exam is a pleasant way to get your feet wet. The test isn't overly complicated or riddled with trick questions; it simply covers the basics of project management. Once you pass the exam and gain confidence in your project management knowledge and skills, you'll be ready to progress to other certifications and be eager to learn the more in-depth topics involved in project management.

What Does This Book Cover?

This book follows that CompTIA Project+ exam blueprint and is divided into chapters that cover major topic areas. Each section is explained in sufficient detail to become a Project+ certified professional. Certain areas have been expanded upon, which are important concepts to know. However, they do not map directly to an exam objective.

You will learn the following information in this book:

Chapter 1: Initiating the Project provides a high-level overview of project management, introducing the basic terminology of project management, including the types of organization structures a project manager may work in.

Chapter 2: Project Team Roles and Responsibilities outlines the various roles involved in project management, from stakeholders and project champion/sponsors to project manager and project team members.

Chapter 3: Creating the Project Charter begins with a detailed discussion of the process groups defined by PMI®: Initiating, Planning, Executing, Monitoring and Controlling, and Closing. It then examines the elements of the project charter, whose definition is the most important Initiating process, and concludes with an overview of the kickoff meeting.

Chapter 4: Creating the Work Breakdown Structure moves into project planning, beginning with documenting the project scope and understanding project influences and constraints and concluding with the decomposition of project tasks into a work breakdown structure.

Chapter 5: Creating the Project Schedule extends planning to creation of a detailed project schedule. It covers the entire schedule planning process, beginning with identifying and sequencing the tasks to be performed and allocating resources. You'll learn how to calculate task durations and the critical path through them, as well as determine milestones and set a baseline and obtain approval. Finally, you'll see how to establish a governance process.

Chapter 6: Resource Planning and Management continues project management into the execution phase with management of the project team. You'll learn about determining resource needs, personnel management, conflict resolution, and the role of the kickoff meeting in team building.

Chapter 7: Defining the Project Budget and Risk Plans covers cost estimating and cost budgeting and the risk activities and strategies for your project. You'll learn the basic techniques of estimating and then tracking costs, along with risk analysis and planning.

Chapter 8: Communicating the Plan covers the role of communication—with stakeholders, team members, and others—in project management. You'll learn what information needs to be communicated and how to do so most effectively.

Chapter 9: Processing Change Requests and Procurement Documents shows how to deal with changing project requirements, with a particular look at the Agile methodology devised for the rapidly evolving requirements of software development. It also looks at the types of organizational change that can affect a project and the specific documents used in dealing with vendors.

Chapter 10: Project Tools and Documentation looks at some of the project tools and documentation needed to inform stakeholders, to document action items and meeting minutes, and to analyze the performance and results of the work of the project. It also reviews the steps a project manager will take in closing out a project.

CompTIA periodically updates and revises exam objectives even after an exam has been released. Even if you have already downloaded the exam objectives, it's always a good idea to check and make sure you have the most current version. To get the exam objectives for the CompTIA Project+ certification, please go to https://certification.comptia .org/certifications/project.

Every question on the Project+ exam is a multiple-choice format. I used this same format for all the questions and practice exams in this study guide.

Many of the examples used to demonstrate practical application of the material in this book focus on IT projects because IT project managers were the original target audience for this exam. However, the techniques and concepts discussed in this book are not limited to IT projects. The information discussed in this book can be applied to projects in any industry.

Interactive Online Learning Environment and Test Bank

I've put together some really great online tools to help you pass the Project+ exam. The interactive online learning environment that accompanies the Project+ exam certification guide provides a test bank and study tools to help you prepare for the exam. By using these tools you can dramatically increase your chances of passing the exam on your first try.

The online section includes the following:

Sample Tests   Many sample tests are provided throughout this book and online, including the assessment test, which you'll find at the end of this introduction, and the chapter tests that include the review questions at the end of each chapter. In addition, there are two bonus practice exams. Use these questions to test your knowledge of the study guide material. The online test bank runs on multiple devices.

Flashcards   The online text bank includes more than 150 flashcards specifically written to hit you hard, so don't get discouraged if you don't ace your way through them at first! They're there to ensure that you're really ready for the exam. And no worries—armed with the review questions, practice exams, and flashcards, you'll be more than prepared when exam day comes! Questions are provided in digital flashcard format (a question followed by a single correct answer). You can use the flashcards to reinforce your learning and provide last-minute test prep before the exam.

Other Study Tools   A glossary of key terms from this book and their definitions are available as a fully searchable PDF.

Go to www.wiley.com/go/sybextestprep to register and gain access to this interactive online learning environment and test bank with study tools.

How to Use This Book

If you want a solid foundation for the serious effort of preparing for the CompTIA Project+ exam, then look no further. I've spent hundreds of hours putting together this book with the sole intention of helping you to pass the exam as well as really learn about the exciting field of project management!

This book is loaded with valuable information, and you will get the most out of your study time if you understand why the book is organized the way it is.

So to maximize your benefit from this book, I recommend the following study method:

1.   Take the assessment test that's provided at the end of this introduction. (The answers are at the end of the test.) It's okay if you don't know any of the answers; that's why you bought this book! Carefully read over the explanations for any question you get wrong and note the chapters in which the material relevant to them is covered. This information should help you plan your study strategy.

2. Study each chapter carefully, making sure you fully understand the information and the test objectives listed at the beginning of each one. Pay extra-close attention to any chapter that includes material covered in questions you missed.

3. Answer all the review questions related to each chapter. (The answers appear in Appendix A.) Note the questions that confuse you, and study the topics they cover again until the concepts are crystal clear. And again—do not just skim these questions! Make sure you fully comprehend the reason for each correct answer. Remember that these will not be the exact questions you will find on the exam, but they're written to help you understand the chapter material and ultimately pass the exam!

4. Try your hand at the practice questions that are exclusive to this book. You can find the questions at www.sybex.com/go/sybextestprep.

5. Test yourself using all the flashcards, which are also found on the download link. These are new and updated flashcards to help you prepare for the Project+ exam and a wonderful study tool.

To learn every bit of the material covered in this book, you'll have to apply yourself regularly and with discipline. Try to set aside the same time period every day to study, and select a comfortable and quiet place to do so. I'm confident that if you work hard, you'll be surprised at how quickly you learn this material.

If you follow these steps and really study in addition to using the review questions, the practice exams, and the electronic flashcards, it would actually be hard to fail the Project+ exam. But understand that studying for the CompTIA exam is a lot like getting in shape—if you do not go to the gym every day, it's not going to happen!

Tips for Taking the Project+ Exam

Here are some general tips for taking your exam successfully:

- Bring two forms of ID with you. One must be a photo ID, such as a driver's license. The other can be a major credit card or a passport. Both forms must have a signature.

- Arrive early at the exam center so you can relax and review your study materials.

- Read the questions carefully. Don't be tempted to jump to an early conclusion. Make sure you know exactly what the question is asking.

- Don't leave any unanswered questions. Unanswered questions are scored against you.

- There will be questions with multiple correct responses. When there is more than one correct answer, there will be a statement at the end of the question instructing you to select the proper number of correct responses, as in "Choose two."

- When answering multiple-choice questions you're not sure about, use a process of elimination to remove the incorrect responses first. This will improve your odds if you need to make an educated guess.

- For the latest pricing on the exam and updates to the registration procedures, refer to the CompTIA site at www.comptia.org.

The Exam Objectives

Behind every certification exam, there are exam objectives. The objectives are competency areas that cover specific topics of project management. The introductory section of each chapter in this book lists the objectives that are discussed in the chapter.

 Exam objectives are subject to change at any time without prior notice and at CompTIA's sole discretion. Please visit the Certification page of CompTIA's website (www.comptia.org) for the most current listing of Project+ exam objectives.

The Project+ exam will test you on four domains, and each domain is worth a certain percentage of the exam. The following is a breakdown of the domains and their representation in the exam:

Domain	% of Examination
1.0 Project Basics	36%
2.0 Project Constraints	17%
3.0 Communication and Change Management	26%
4.0 Project Tools and Documentation	21%
Total	**100%**

Project+ Exam Map

The following objective map will allow you to find the chapter in this book that covers each objective for the exam.

1.0 Project Basics

Exam Objective	Chapter
1.1 Summarize the properties of a project.	**1**
Temporary	1
Start and finish	1
Unique	1
Reason/purpose	1
Project as part of a program	1
Project as part of a portfolio	1
1.2 Classify project roles and responsibilities.	**2**
Sponsor/champion	2
Project manager	2
Project coordinator	2
Stakeholder	2
Scheduler	2
Project team	2
Project Management Office (PMO)	2

2.0 Project Constraints

3.0 Communication and Change Management

Exam Objective	Chapter
Tailor method based on content of message.	8
Criticality factors	8
Specific stakeholder communication requirements	8
3.3 Explain common communication triggers and determine the target audience and rationale.	**8**
Audits	8
Project planning	8
Project change	8
Risk register updates	8
Milestones	8
Schedule changes	8
Task initiation/completion	8
Stakeholder changes	8
Gate reviews	8
Business continuity response	8
Incident response	8
Resource changes	8
3.4 Given a scenario, use the following change control process within the context of a project.	**9**
Change control process	9
Types of common project changes	9
3.5 Recognize types of organizational change.	**9**
Business merger/acquisition	9

4.0 Project Tools & Documentation

Assessment Test

1. Which of these terms describes a critical path task?
   A. Hammock
   B. Zero float
   C. Critical task
   D. Mandatory task

2. Resource management concepts include several categories of resources. All of the following are a type of resource except for which one?
   A. Shared resource
   B. Remote and in-house
   C. Dedicated resources
   D. Benched resources

3. This tool is often used in the vendor selection process to pick a winning bidder.
   A. Weighted scoring model
   B. Bidder conference
   C. RFQ
   D. SOW

4. From the following list of options, select three of the five common stages of development that project teams progress through. Choose three.
   A. Forming
   B. Acquiring
   C. Storming
   D. Adjourning
   E. Negotiating
   F. Norming
   G. Compromising

5. In this organizational structure, you report to the director of project management, and your team members report to their areas of responsibility (accounting, human resources, and IT). You will have complete control of the project team members' time and assignments once the project is underway. Which type of organization does this describe?
   A. Projectized
   B. Functional
   C. Hierarchical
   D. Matrix

6. This describes how you will know the completed deliverables are satisfactory.

   A. Acceptance criteria

   B. KPIs

   C. Metrics

   D. EVM

7. Which of the following describes the responsibilities of the project sponsor?

   A. Provides or obtains financial resources

   B. Monitors the delivery of major milestones

   C. Runs interference and removes roadblocks

   D. Provides the project manager with authority to manage the project

   E. All of the above

8. Project managers may spend up to 90 percent of their time doing which of the following?

   A. Interacting with the project stakeholders

   B. Interacting with the project sponsor

   C. Interacting with the project team members

   D. Communicating

9. This is a temporary way of resolving conflict and is considered a lose-lose technique. It emphasizes the areas of agreement over the areas of disagreement.

   A. Smoothing

   B. Forcing

   C. Avoiding

   D. Confronting

10. Which of these is not an example of a project selection method?

    A. Cost-benefit analysis

    B. Expert judgment

    C. Top-down estimating

    D. Scoring model

11. You're the project manager on a project where the scope has expanded. The change has been approved by the change control board (CCB). What steps must you take to acknowledge the new scope? Choose two.

    A. Update the project management plan.

    B. Update the SOW.

    C. Communicate the change to stakeholders and team members.

    D. Submit a change request.

    E. Log the request on the change request log.

12. Some team members on your team are driving each other crazy. They have different ways of organizing the tasks they are both assigned to, and the disparity in styles is causing them to bicker. Which of the following describes this situation?

   A. This is a constraint that's bringing about conflict on the team.

   B. This should be escalated to the project sponsor.

   C. This is a common cause of conflict.

   D. This is a team formation stage that will pass as they get to know each other better.

13. Fishbone diagrams, Pareto diagrams, process diagrams, Gantt charts, and run charts are examples of which of the following?

   A. Examples of various project management tools used to plan the work of the project

   B. Examples of various project management tools used to monitor and control project work

   C. Examples of various scope management tools to control the quality of deliverables

   D. Examples of various scope management tools used to manage and control scope creep

14. Which of the following project documents created in the project Closing process group describes what went well and what didn't go well on the project?

   A. Project close report

   B. Postmortem

   C. Lessons learned

   D. Post-project review

15. This document authorizes the project to begin.

   A. Project request

   B. Project concept document

   C. Project charter

   D. Project scope statement

16. Which of these can convey that you've achieved the completion of an interim key deliverable?

   A. Completion criteria

   B. Milestone

   C. Gantt chart

   D. Project sign-off document

17. All of the following are factors that influence communication methods except which one?

   A. Language barriers

   B. Technological barriers

   C. Task completion

   D. Cultural differences

E. Intraorganizational differences

F. A, B, D

G. All of the above

18. This person is responsible for removing obstacles so the team can perform their work, assisting the product owner in defining backlog items, and educating the team on Agile processes.

A. Project manager

B. Scrum master

C. Functional manager

D. Subject-matter expert

19. This project management methodology uses self-organized, self-directed teams; it uses an iterative approach; and it is highly interactive.

A. PRINCE2

B. PMI®

C. Agile

D. Waterfall

20. This is often added to the project schedule to determine whether the work is correct.

A. Approval gate

B. Quality gate

C. Milestone gate

D. Governance gate

21. This is the final, approved version of the project schedule. All of the following are true regarding this term except for which of the following?

A. It will prevent future schedule risk.

B. It's approved by the stakeholders, sponsor, and functional managers.

C. It's used to monitor project progress throughout the remainder of the project.

D. This describes a schedule baseline.

22. A well-written change control process should include which of the following components? Choose two.

A. A description of the type of change requested

B. The amount of time the change will take to implement

C. The cost of the change

D. How to update the affected project planning documents

E. The stages at which changes are accepted

23. Which cost-estimating technique relies on estimating work packages and then rolling up these estimates to come up with a total cost estimate?

    A. Top-down

    B. Parametric

    C. Bottom-up

    D. Analogous

24. In project management, the process of taking high-level project requirements and breaking them down into the tasks that will generate the deliverables is called what?

    A. Analyzing

    B. Decomposing

    C. Process flow diagram

    D. Documenting

25. Who is responsible for assembling the project's team members?

    A. Project sponsor

    B. Project stakeholders

    C. Project customer

    D. Project manager

26. All of the following represent a category of contract most commonly used in procurement except for which one?

    A. Time-and-materials

    B. Cost-reimbursable

    C. Fixed-price

    D. Requests for proposal

27. This is a deliverables-oriented hierarchy that defines the work of the project.

    A. Scope document

    B. Work breakdown structure

    C. Scope management plan

    D. Project plan

28. Your project sponsor has expressed their need to have real-time project information at their fingertips. Which of the following is the best way to meet this need?

    A. By creating a project dashboard with scope, cost, and time elements

    B. By updating the project status report on a daily basis

    C. By sending an email every morning to the sponsor describing the current project status

    D. By meeting face to face with the sponsor every day to update them on status

29. Your project sponsor told you that the due date for the project is a key to its success and there is no chance of the date changing. What is this known as?

A. A constraint

B. An influence

C. A deliverable

D. A management directive

30. Risk analysis includes all of the following except for which one?

A. Identifying risk

B. Determining a risk response plan

C. Determining an order-of-magnitude estimate for responses

D. Determining probability and impact and assigning a risk score

31. Which of the following are project management phases? Choose three.

A. Scheduling

B. Planning

C. Executing

D. Communicating

E. Documenting

F. Budgeting

G. Closing

32. When taking over an incomplete project, what item should be of most interest to the new project manager?

A. Project concept statement

B. Project charter

C. Project scope statement

D. Project plan

33. This chart or diagram is a type of histogram that rank-orders data by frequency over time.

A. Run chart

B. Scatter diagram

C. Fishbone diagram

D. Pareto chart

34. These performance measurements are defined when you create the project management plan and are monitored and tracked during the Monitoring and Controlling phase to determine whether the project is meeting its goals.

A. SPI

B. Balanced score card

C. CPI

D. KPI

35. What key meeting is held after the project charter is signed and/or at the beginning of the Executing process?

    A. Project kickoff

    B. Project review

    C. Project overview

    D. Project status meeting

36. What is the best way to prevent scope creep?

    A. Make sure the requirements are thoroughly defined and documented.

    B. Put a statement in the charter that no additions to the project will be allowed once it's underway.

    C. Alert the sponsor that you will not be taking any change requests after the project starts.

    D. Inform stakeholders when they sign the project scope statement that no changes will be accepted after the scope statement is published.

37. Which of these statements describes an assumption?

    A. Our senior web developer will be available to work on this project.

    B. The electrical capacity at the site of the project event may not be adequate.

    C. The project's due date is June 27.

    D. There's a potential for the server administrator to receive a promotion during the course of this project.

38. Your subject-matter expert tells you that her most likely estimate to complete her task is 40 hours. The task starts on Thursday, January 20, at 8 a.m. The team works 8-hour days, and they do not work weekends. Which day will the task end?

    A. January 24

    B. January 25

    C. January 27

    D. January 26

39. You are working for an organization and just learned that another organization with more power and influence is taking over your organization at the first of the year. What does this describe?

    A. Business merger

    B. Business demerger

    C. Business venture

    D. Business acquisition

40. What are the two types of charts that you might utilize to display the project schedule? Choose two.

    A. Run chart

    B. Gantt chart

    C. Milestone chart

    D. CPM

    E. Histogram

41. All of the following describe types of project endings except for which one?

    A. Integration

    B. Starvation

    C. Addition

    D. Extinction

    E. Attrition

42. Demonstrating competency, respect, honesty, integrity, openness, and doing what you say you'll do is an example of which of the following?

    A. Team building

    B. Managing team resources

    C. Demonstrating leadership skills

    D. Trust building

43. This is the approved, expected cost of the project.

    A. Expenditure budget

    B. Expenditure baseline

    C. Cost budget

    D. Cost baseline

44. "Install an Interactive Voice Response System that will increase customer response time by an average of 15 seconds and decrease the number of customer service interactions by 30 percent" is an example of which of the following elements of the project charter?

    A. High-level requirements

    B. Goals and objectives

    C. Project description

    D. Milestone

45. The network communication model is a visual depiction of what?

    A. Lines of communication

    B. Participant model

    C. Communication model

    D. Participant communication model

46. This estimating technique uses the most likely, optimistic, and pessimistic estimates to come up with an average cost or duration estimate.

    A. Analogous estimate

    B. Bottom-up estimate

    C. Parametric estimate

    D. Three-point estimate

47. In this type of organization, the project manager shares responsibility for team member assignments and performance evaluations with the functional manager.

    A. Functional

    B. Projectized

    C. Hierarchical

    D. Matrix

48. Which of the following is not true regarding cost estimating?

    A. Cost estimates are provided by team members.

    B. Cost estimate accuracy depends on the technique used to determine the estimate.

    C. Cost estimates have a quality factor built into them.

    D. Cost estimates are inputs to the project budget and used to determine the total project cost.

49. You're the project manager for a small project that is in the Closing phase. You prepare closure documents and take them to the project sponsor for sign-off. The project sponsor says that the documents are not needed because the project is so small. What should you tell the sponsor?

    A. You're sorry to have bothered them and will close the project without sign-off.

    B. The sponsor is the one who needs to sign off on the documents, showing that the project is officially closed.

    C. You offer to have a stakeholder sign off in the sponsor's place.

    D. You offer to sign off on the documents yourself.

50. This meeting is held so that team members can answer three questions: what work they completed yesterday, what work they will complete today, and what obstacles stand in their way.

    A. This describes a sprint planning meeting, which is part of the Agile methodology.

    B. This describes a Scrum meeting, which is part of the Agile methodology.

    C. This describes a daily standup, which is part of the waterfall methodology.

    D. This describes a retrospective meeting, which is part of the PMI® methodology.

Answers to Assessment Test

1. B. Tasks with zero float are critical path tasks, and if delayed, they will cause the delay of the project completion date. For more information, please see Chapter 5.

2. B. Remote and in-house resources are categorized as personnel management activities in the CompTIA objectives. For more information, please see Chapter 6.

3. A. A weighted scoring model is a tool that weights evaluation criteria and provides a way to score vendor responses. Bidder conferences, IFB, and SOW are all used during vendor solicitation. For more information, please see Chapter 9.

4. A, C, F. The five stages of team development are forming, storming, norming, performing, and adjourning. For more information, please see Chapter 6.

5. A. This describes a projectized organization because the project manager works in a division whose sole responsibility is project management, and once the team members are assigned to the project, the project manager has the authority to hold them accountable to their tasks and activities. For more information, please see Chapter 1.

6. A. Acceptance criteria describe how to determine whether the deliverables are complete and meet the requirements of the project. For more information, please see Chapter 4.

7. E. A project sponsor is responsible for obtaining financial resources for the project, monitoring the progress of the project, and handling escalations from the project manager. For more information, please see Chapter 2.

8. D. Project managers may spend up to 90 percent of their time communicating. For more information, please see Chapter 1.

9. A. Smoothing is a lose-lose conflict-resolution technique. It is a temporary way to resolve conflict. Avoiding can also be a lose-lose conflict technique, but it isn't temporary in nature and doesn't emphasize anything because one of the parties leaves the discussions. For more information, please see Chapter 6.

10. C. Cost-benefit analysis, expert judgment, and scoring model are all project selection techniques. Top-down estimating is a type of cost estimating. For more information, please see Chapter 1.

11. A, C. Any time there's a significant change to the project, the project management plan must be updated and the stakeholders notified of the change. Options D and E would have been done before the approval by the CCB. Options A and C occur after an approval. For more information, please see Chapter 9.

12. C. This situation describes varying work styles that are a common cause of conflict. Competing resource demands and constraints are also common causes of conflict. Conflicts are anything that restricts or dictates the actions of the project team. And issues like this should almost never have to be escalated to the project sponsor. For more information, please see Chapter 6.

13. B. The tools described in this question are used during the Monitoring and Controlling phase of the project to monitor project work and assure it meets expectations. It also helps in determining corrective actions needed to get the project back on track. For more information, please see Chapter 10.

14. C. Lessons learned describe what went well and what didn't go well on the project. Lessons learned are included in the project close report, the postmortem report, and the post-project review. For more information, please see Chapter 10.

15. C. The project charter authorizes the project to begin. For more information, please see Chapter 3.

16. B. Milestones often signal that you've completed one of the key deliverables on the project. For more information, please see Chapter 5.

17. F. Task completion is a communication trigger. The remaining options are examples of factors that influence communications. For more information, please see Chapter 8.

18. B. The Scrum master is responsible for removing obstacles that are getting in the way of the team performing the work. They work with the product owner to help define backlog items, and they educate team members on the Agile process. For more information, please see Chapter 9.

19. C. The Agile project management methodology uses self-organized and self-directed teams. The other options don't use these types of teams. For more information, please see Chapter 9.

20. B. A quality gate is added to the schedule as a checkpoint to determine whether the work meets quality standards. For more information, please see Chapter 5.

21. A. The schedule baseline is the final, approved version of the schedule and is signed by the stakeholders, sponsor, and functional managers. Having a schedule baseline will not prevent future schedule risk. For more information, please see Chapter 5.

22. A, D. The amount of time and money a change will require are outcomes of a change control process, not inputs to the process. For more information, please see Chapter 9.

23. C. The bottom-up cost-estimating method is the most precise because you begin your estimating at the activities in the work package and roll them up for a total estimate. For more information, please see Chapter 7.

24. B. Decomposition is the process of analyzing the requirements of the project in such a way that you reduce the requirements down to the steps and tasks needed to produce them. For more information, please see Chapter 4.

25. D. The project manager assembles the team members for the project. The project manager may get input from the sponsor, stakeholders, or customers, but it is the project manager who decides what the formation of the team should be. For more information, please see Chapter 1.

26. D. The three categories of contracts most often used to procure goods and services are time-and-materials, cost-reimbursable, and fixed-price. Requests for proposal are not contracts. For more information, please see Chapter 9.

27. B. The WBS is a deliverables-oriented hierarchy that defines all the project work and is completed after the scope management plan and scope statement are completed. For more information, please see Chapter 5.

28. A. The best way to provide this information is to create a dashboard that provides real-time, updated information in a succinct and easy-to-read format. For more information, please see Chapter 10.

29. A. This describes a constraint. Constraints dictate or restrict the actions of the project team. For more information, please see Chapter 4.

30. C. Determining an order-of-magnitude estimate is used for cost or duration estimating, not risk analysis. For more information, please see Chapter 7.

31. B, C, G. Initiating, Planning, Executing, Monitoring and Controlling, and Closing are the five project management phases or process groups. For more information, please see Chapter 3.

32. C. The project's scope statement should be of most interest to the new project manager. The scope statement describes the product description, key deliverables, success and acceptance criteria, exclusions, assumptions, and constraints. For more information, please see Chapter 4.

33. D. A Pareto diagram rank-orders data by frequency over time. For more information, please see Chapter 10.

34. D. Key performance indicators (KPIs) are measurable elements of project success defined when you create the project management plan and measured and monitored throughout the Monitoring and Controlling process. For more information, please see Chapter 10.

35. A. The project kickoff meeting is held after the project charter is signed and at the beginning of the Executing process. It serves to introduce team members, review the goals and objectives of the project, review stakeholder expectations, and review roles and responsibilities for team members. For more information, please see Chapter 6.

36. A. The best way to avoid scope creep is to make sure the project's requirements have been thoroughly defined and documented. For more information, please see Chapter 4.

37. A. Assumptions are those things we believe to be true for planning purposes. Options B and D describe risks, while option C describes a constraint. For more information, please see Chapter 3.

38. D. The task begins on January 20, which is day 1. The team does not work weekends, so the completion date, based on an eight-hour workday, is January 26. For more information, please see Chapter 5.

39. D. This question describes a business acquisition. Companies that are acquiring others have the power and influence to make decisions. A business merger is a mutually agreeable arrangement where power is shared among the entities. For more information, please see Chapter 9.

40. B, C. Gantt charts and milestone charts are the most commonly used formats to display a project schedule. For more information, please see Chapter 5.

41. E. Integration occurs when resources are distributed to other areas of the organization, and addition occurs when projects evolve into ongoing operations. Starvation is a project ending caused by resources being cut off from the project. Extinction occurs when the project work is completed and is accepted by the stakeholders. For more information, please see Chapter 10.

42. D. This is an example of trust building. As a project manager, you must do what you say you'll do and demonstrate the traits stated in the question. For more information, please see Chapter 6.

43. D. The cost baseline is the approved, expected cost of the project. For more information, please see Chapter 7.

44. B. Goals and objectives are specific and measurable. Project descriptions describe the key characteristics of the product, service, or result of the project. These are characteristics, but the clue in this question is the quantifiable results you're looking for at the conclusion of the project. The project description describes the project as a whole, and milestones describe major deliverables or accomplishments for the project. For more information, please see Chapter 3.

45. A. Lines of communication describe how many lines of communication exist between participants. The network communication model is a visual depiction of the lines of communication. For more information, please see Chapter 8.

46. D. The three-point estimating technique averages the most likely, optimistic, and pessimistic estimates to determine an overall cost or duration estimate. For more information, please see Chapter 7.

47. D. Project managers share authority with functional managers in a matrix organization. For more information, please see Chapter 2.

48. B. Cost estimates do not have a quality factor built into them. They are provided by team members, and the accuracy of the estimate depends on the estimating technique used. All the estimates are used as inputs to the budget to come up with the total project cost. For more information, please see Chapter 7.

49. B. The sponsor is the one who must sign off on the completion of the project, whether successful or unsuccessful. Just as the sponsor is authorized to expend resources to bring forth the project's deliverables, the sponsor must also close the project and sign off. For more information, please see Chapter 10.

50. B. This describes a Scrum meeting, also known as a daily standup meeting, which is part of the Agile methodology. For more information, please see Chapter 9.

Chapter

1

Initiating the Project

THE COMPTIA PROJECT+ EXAM TOPICS COVERED IN THIS CHAPTER INCLUDE

✓ **1.1 Summarize the properties of a project.**

- Temporary

- Start and finish

- Unique

- Reason/purpose

- Project as part of a program

- Project as part of a portfolio

✓ **1.5 Identify common project team organizational structures.**

- Functional

  - Resources reporting to functional manager

  - Project manager has limited or no authority.

- Matrix

  - Authority is shared between functional managers and project managers.

  - Resources assigned from Functional area to project

  - Project manager authority ranges from weak to strong.

  - Projectized

  - Project manager has full authority.

  - Resources report to project manager.

  - Ad hoc resources

Your decision to take the CompTIA Project+ exam is an important step in your career aspirations. Certification is important for project managers because many employers look for project management certification in addition to real-life experience and evidence of formal education from job applicants. This book is designed to provide you with the necessary concepts to prepare for the Project+ exam. Much of the information here will be based on the Knowledge Areas documented in *A Guide to the Project Management Body of Knowledge (PMBOK Guide®)* published by the Project Management Institute (PMI®). The book will include tips on how to prepare for the exam, as well as examples and real-world scenarios to illustrate the concepts.

This chapter will cover the definitions and characteristics of a project, provide a high-level overview of project management, describe the difference between a program and a portfolio, and explain how organizations are structured.

Defining the Project

Projects exist to bring about or fulfill the goals of the organization. Most projects benefit from the application of a set of processes and standards known as *project management*. Let's start with some fundamental questions.

- What makes a new assignment a project?

- How do you know if you are working on a project?

- What distinguishes a project from an operational activity?

Projects involve a team of people, and so do day-to-day business activities. They both involve following a process or a plan, and they both result in activities that help reach a goal. So, what is so different about a project? Let's explore all of these questions in the following sections.

Identifying the Project

A *project* is a temporary endeavor that has definite beginning and ending dates, and it results in a unique product, service, or result. A project is considered a success when the goals it sets out to accomplish are fulfilled and the stakeholders are satisfied with the results.

Projects also bring about a product, service, or result that never existed before. This may include creating tangible goods, implementing software, writing a book, planning and executing an employee appreciation event, constructing a building, and more. There is no limit to what can be considered a project as long as it fits the following criteria:

Unique   A project is typically undertaken to meet a specific business objective. It involves doing something new, which means that the end result should be a unique product or service. These products may be marketed to others, may be used internally, may provide support for ongoing operations, and so on.

Temporary   Projects have definite start and end dates. The time it takes to complete the work of the project can vary in overall length from a few weeks to several years, but there is always a start date and an end date.

Reason or Purpose   A project comes about to fulfill a purpose. This might include introducing a new product, fulfilling a business objective or strategic goal, satisfying a social need, and any number of other reasons. It's important to document and communicate the purpose and reasons for the project so that team members remain focused on achieving the goals of the project.

Stakeholder Satisfaction   A project starts once it's been identified, the objectives have been outlined in the project charter, and appropriate stakeholders have approved the project plan. A project ends when those goals have been met to the satisfaction of the stakeholders.

Once you've identified the project, you'll validate the project (we'll cover this topic in the section "Validating the Project" later in this chapter) and then write the project charter and obtain approval for the charter. We'll talk in more detail about the project charter in Chapter 3, "Creating the Project Charter."

Programs and Portfolios

Projects are sometimes managed as part of a program or portfolio. A *program* is a group of related projects that are managed together using coordinated processes and techniques. The collective management of a group of projects can bring about benefits that wouldn't be achievable if the projects were managed separately.

Portfolios are collections of programs, subportfolios, and projects that support strategic business goals or objectives. Unlike programs, portfolios may consist of projects that are not related.

Here's an example to help clarify the difference between programs and portfolios. Let's say your company is in the construction business. The organization has several business units: retail construction, single-family residential buildings, and multifamily residential buildings. Individually, each of the business units may comprise a program. For example, retail construction is a program because all the projects within this program exist to create new retail-oriented buildings. This is not the same as single-family home construction (a different program), which is not the same as multifamily residential construction (a different program). Collectively, the programs and projects within all of these business units make up the portfolio. Other projects and programs may exist within this portfolio as well, such as parking structures, landscaping, and so on.

Programs and projects within a portfolio are not necessarily related to one another in a direct way. And projects may independently exist within the portfolio (in other words, the project isn't related to a program but belongs to the portfolio). However, the overall objective of any program or project in a portfolio is to meet the strategic objectives of the portfolio, which in turn should meet the strategic objectives of the business unit or corporation.

Understanding Operations

Operations are ongoing and repetitive. They don't have a beginning date or an ending date, unless you're starting a new operation or retiring an old one. Operations typically involve ongoing functions that support the production of goods or services. Projects, on the other hand, come about to meet a specific, unique result and then conclude.

It's important to understand that projects and operations go hand in hand in many cases. For example, perhaps you've been assigned to research and implement state-of-the-art equipment for a shoe manufacturing plant. Once the implementation of the equipment is complete, the project is concluded. A handoff to the operations team occurs, and the everyday tasks the equipment performs become an ongoing operation.

Don't be confused by the term *service* regarding the definition of a project. Providing janitorial services on a contract is operations; providing contract Java programmers for 18 months to work on an IT project is a project.

Let's look at the definition of two more terms. *Project management* brings together a set of tools and techniques—performed by people—to describe, organize, and monitor the work of project activities. *Project managers* (PMs) are the people responsible for applying these tools to the various project activities. Their primary purpose is to integrate all the components of the project and bring it to a successful conclusion. Managing a project involves many skills, including dealing with competing needs for your resources, obtaining adequate budget dollars, identifying risks, managing to the project requirements, interacting with stakeholders, staying on schedule, and ensuring a quality product.

We'll spend the remainder of this book describing the tools and techniques you'll use to accomplish the goals of the project, including the key concepts you'll need to know for the exam. Many of the standards surrounding these techniques are documented in the *PMBOK Guide®*.

Using the *PMBOK Guide®*

Project management standards are documented in *A Guide to the Project Management Body of Knowledge* (*PMBOK Guide®*), published by the Project Management Institute. PMI® sets the global de facto standard in project management. It's a large organization with more than 700,000 members from countries around the globe.

In addition to publishing the *PMBOK Guide®*, PMI® also manages two rigorous certification exams for individual project managers: the Certified Associate in Project Management (CAPM)® and the Project Management Professional (PMP)®. The *PMBOK Guide®* is the basis for the exam portion of the CAPM and PMP certifications. If you continue in a career in project management, you may decide to study and sit for the CAPM or PMP certification exams. The material you will study to prepare for the Project+ exam is an excellent foundation on which to build your project management knowledge.

Understanding Organizational Structures

The structure of your organization has an impact on many aspects of project management, including the authority of the project manager and the process to assign resources.

Project managers are often frustrated by what appear to be roadblocks in moving the project forward, but in many cases, the root issue is the organizational structure itself and how it operates. The following sections will cover the different types of organizational structures and how they influence the way projects are conducted.

The Functional Organization

The classic organizational structure is the *functional organization*, as shown in Figure 1.1. In this structure, the staff is organized along departmental lines, such as IT, marketing, sales, network, human resources, public relations, customer support, and legal. Each department is managed independently with a limited span of control. This organizational type is hierarchical, with each staff member reporting to one supervisor, who in turn reports to one supervisor, and so on up the chain. Figure 1.1 shows a typical functional organization.

FIGURE 1.1 The functional organization

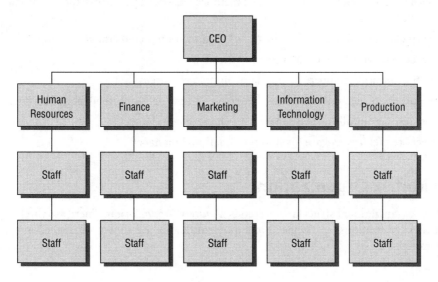

A functional organization often goes about the work of the project in a siloed fashion. That is, the project deliverables are worked on independently in different departments. This can cause frustration among project managers, because they are the ones held accountable for the results of the project, but they have no means of holding team members from other departments accountable for completing project deliverables.

A project manager in a functional organization should develop strong working relationships with the functional managers. Functional managers are responsible for assigning work to the employees who report to them. They are also responsible for rating the performance of the employees and determining their raises or bonuses. This, as you can imagine, sets up a strong loyalty between the employee and the functional manager as opposed to the employee and the project manager. However, that doesn't mean project managers can't be successful in this type of organization. Building a relationship with the functional managers and maintaining open communications is the key to successful projects in this type of structure. It also helps a great deal if you can contribute to the employee's performance ratings by rating their work on the project.

Project managers have little formal authority in this type of structure, but it doesn't mean their projects are predestined for failure. Communication skills, negotiation skills, and strong interpersonal skills will help assure your success in working within this type of environment.

The functional organization is the most common organizational structure and has endured for centuries. The advantages of a functional organization include the following:

- Growth potential and a career path for employees

- The opportunity for those with unique skills to flourish

- A clear chain of command (each staff member has one supervisor—the functional manager)

The typical disadvantages of a functional organization include the following:

- Project managers have limited to no authority.

- Multiple projects compete for the same limited resources.

- Resources are generally committed part-time to the project rather than full-time.

- Issue resolution follows the department chain of command.

- Project team members are loyal to the functional manager.

The Matrix Organization

The next organizational structure covered is a *matrix organization*. There are three types of matrix organizations, as discussed in a moment. Figure 1.2 shows a balanced matrix organization.

FIGURE 1.2   The balanced matrix organization

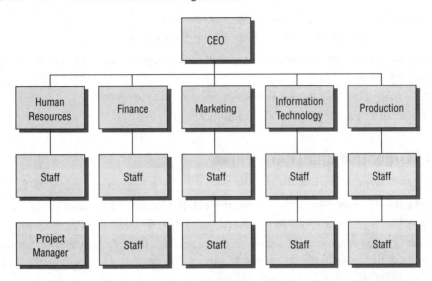

Matrix organizations typically are organized along departmental lines, like a functional organization, but resources assigned to a project are accountable to the project manager for all work associated with the project. The project manager is often a peer of the functional staff managers. The team members working on the project often have two or more supervisors— their functional manager and the project manager (or managers) they are reporting to.

Project managers working in a matrix organization need to be clear with both the project team members and their respective functional managers about assignments and results regarding the following:

- Those outcomes for which the team member is accountable to the project manager
- Those outcomes for which the team member is accountable to the functional manager

The team member should be accountable to only one person for any given outcome so as to avoid confusion and conflicting direction.

Another trouble area in a matrix organization is availability of resources. If you have a resource assigned 50 percent of the time to your project, it's critical that the functional manager, or other project managers working with this resource, is aware of the time commitment this resource has allocated to your project. If time-constraint issues like this are not addressed, project managers may well discover they have fewer human resources for the project than first anticipated. Addressing resource commitments at the beginning of the project, both during preproject setup and again during the planning phase, will help prevent problems down the road.

In a typical matrix organization, functional managers assign employees to the project, while project managers assign tasks associated with the project to the employee.

The following are the typical characteristics of a matrix organization:

- Project manager authority ranges from weak to strong.
- There is a mix of full-time and part-time project resources.
- Resources are assigned to the project within their respective functional areas by a functional manager.
- Project managers and functional managers share authority levels.
- There is better interdepartmental communication.

Matrix Organizations Times Three

There are three types of matrix organizations:

Strong Matrix   The strong matrix organization emphasizes project work over functional duties. The project manager has the majority of power in this type of organization.

Weak Matrix   The weak matrix organization emphasizes functional work over project work and operates more like a functional hierarchy. The functional managers have the majority of power in this type of organization.

Balanced Matrix   A balanced matrix organization shares equal emphasis between projects and functional work. Both the project manager and the functional manager share power in this type of structure.

Matrix organizations allow project managers to focus on the work of the project. The project team members, once assigned to a project, are free to focus on the project objectives with minimal distractions from the functional department.

It is important that you understand what type of matrix organization you're working in because the organizational type dictates the level of authority you'll have. But don't be fooled into thinking you will have your way more easily in a strong matrix environment. It is still essential that you keep the lines of communication open with functional managers and inform them of status, employee performance, future needs, project progress, and so on.

The Projectized Organization

The last type of organizational structure covered is the *projectized organization*, which is shown in Figure 1.3. This organizational structure is far less common than the other two discussed. In this environment, the focus of the organization is projects, rather than functional work units.

FIGURE 1.3   The projectized organization

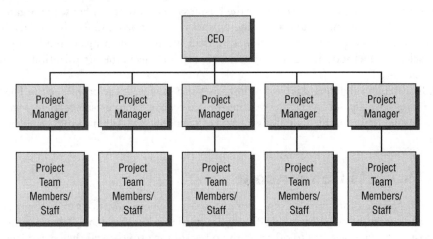

Project managers have the majority of power in this type of structure. They are responsible for making decisions regarding the project and for acquiring and assigning resources from inside or outside the organization. Support staff, such as human resources, administrative support, accounting, and so on, often report to the project manager in a projectized environment.

One of the advantages of this type of organization is that team members are *colocated*, meaning they work together at the same physical location. Other advantages of this structure include the following:

- Project manager has full authority to manage the project and resources.
- Full-time resources are assigned to the project and report to the project manager.
- Loyalty is established with the project manager.
- Other ad hoc resources may report to the project manager.
- There is dedicated project support staff.

One of the biggest drawbacks of a projectized organization is reassigning project team members once the project ends. There may not always be a new project waiting for these resources. Again, it's essential that communication is occurring among project managers across the organization so that the complex timing of increasing or decreasing resources is managed as efficiently as possible.

Validating the Project

Stakeholders have many reasons for bringing about a project. Most organizations don't have the resources or time to execute every project that every stakeholder would like implemented. Validating the project involves two steps: preparing the business case and identifying and analyzing the project stakeholders. But there are some steps along the way you need to understand before and after the business case is written. First, the organization needs to have an understanding of the business need or demand for the project. Then, the business case is created, which includes a justification for the project, and finally, project selection methods are used to determine which projects the organization should implement.

The first step in validating a project is preparing and validating the business case. The business case typically documents the reasons the project came into existence. Before we dive into the business case specifics, we'll cover some of the needs and demands that bring about projects.

How Projects Come About

Projects come about for many reasons. For example, some organizations exist to generate profits, and many create projects specifically designed to meet this goal. Other organizations exist to provide services to others with no regard for profits. And they may bring about projects to enhance their ability to meet the demand for their services. No matter what the reason for bringing about a project, most of them will fall into one of the seven needs or demands described next:

Market Demand   The demands of the marketplace can drive the need for a project. For example, the proliferation of handheld devices has created a need for rechargeable batteries that are capable of holding a charge for 12 hours or more.

Strategic Opportunity/Business Need   Business needs often drive projects that involve information technology solutions. For example, an organization's accounting system is outdated and no longer able to keep up with current technology. A new system is implemented to help the organization become more efficient and create reports in a timelier manner.

Customer Request   Customer requests can generate an endless supply of potential projects. For example, perhaps the discussions at a recent customer focus group brought about the idea for a new product offering.

Technological Advance   Technology and business needs sometimes strike us as a chicken-and-egg scenario. Is it the technology that drives the business to think it needs a new product or service, or does the business need drive the development of the new technology? Both scenarios exist, and both bring about the need for new projects.

Legal Requirement   Local, state, and federal regulations change during every legislative session and may drive the need for a new project. For example, a city may pass an ordinance allowing photos of red-light violations at busy intersections. The new equipment must then be procured and installed. Federal regulations requiring the encryption and secure storage of private data may bring about the need for a project to fulfill these requirements.

Real World Scenario

Assessing the Impact of Regulations and Legal Requirements

Projects often have legislative, regulatory, or other third-party restrictions imposed upon their processes or outputs. For example, suppose you are managing a project that will create a new information technology system for a funds management company, one that's in the business of managing individual stock portfolios. You can imagine that this company is heavily regulated by the Securities and Exchange Commission (SEC) and that your new system, in turn, will encounter several regulatory guidelines that you must follow. The security aspect of your new system is especially pertinent. You must be able to assure the SEC and your shareholders that the system is secure.

It's important that a project manager be able to not only recognize the need to investigate specific industry regulations and requirements but also communicate this need and its associated impact on the project scope and project plan to the stakeholders. Here are a few examples of the many external considerations you need to account for when implementing a technology-based project:

Legal and Regulatory Conditions   Know the statutes covering the type of activity your deliverable involves. For example, if you collect information about customers, make certain you are complying with privacy laws. Also, you may face government reporting and documentation requirements or public-disclosure rules.

Licensing Terms   Suppose that part of your project requires that developers write some programming code according to a Microsoft application programming interface (API). You need to be well aware of the licensing ramifications associated with using a Microsoft API. Trademark, copyright, and intellectual property issues all enter into this category.

Industry Standards   Industry standards exist in almost every aspect of business. Pharmaceutical companies, car manufacturers, food services, and so on, all have industry standards that describe best practices for preparing, manufacturing, shipping, and any number of other elements of their business.

Considerations for industry standards in your organization must be accounted for in the project plan and budget.

Environmental Considerations   Many organizations today are actively involved in green efforts to protect the environment. For example, perhaps a new Environmental Protection Agency (EPA) mandate requires extra equipment and processes to be implemented in your production assembly line to minimize pollution output. Therefore, a project is required.

Social Need   Social needs or demands can bring about projects in a variety of ways. For example, a small developing country may have the need for safe, clean drinking water, so a project is initiated to purchase and install a new filtering system. Another example may include bringing about a project to develop a vaccine for a new flu virus that's predicted to hit the nation.

The needs or demands that bring about a project are usually stated in the business case. You'll look at the business case next.

Business Case Validation and Stakeholder Identification

A project is validated by preparing a business case and identifying and analyzing stakeholders. There are several steps to validating a business case as well. You'll learn about all these project validation steps next.

Validating the Business Case

One of the first things a project manager can do at the onset of a project is to understand the business reason for the project. The business case, which is often based on one of the needs or demands discussed in the previous section, is a written document or report that helps executive management and key stakeholders determine the benefits and rewards of the project. It documents the business need or justification for the project and will often include high-level details regarding estimated budgets and timelines for completing the project.

Justification   Justification describes the benefits to the organization for undertaking the project. These benefits can include tangible and intangible benefits and should include the reasons for bringing about the project. Justification can be a section within the business case or an independent document.

Alignment to the Strategic Plan   Alignment to the strategic plan can also be included within the business case, and it should describe how the project and its outcomes will align to the organization's overall strategic plan. If the reason for the project doesn't support the strategic plan, there's really no reason to undertake the project.

Alternative Solutions   This should include a high-level description of costs, the feasibility of implementing each alternative, the expected results of each alternative solution, and a description of any impacts to the organization as a result of this solution. (Cost-benefit, payback, and other financial analyses are generally included in this section of the business case.)

Recommended Solution   This section details the recommended solution.

Feasibility Study   A *feasibility study* is undertaken for several reasons. Feasibility studies can determine whether the project is doable and likely to succeed. They examine the viability of the product, service, or result of the project. They may also examine technical issues related to the project and determine whether it's feasible, reliable, and easily assimilated into the organization's existing infrastructure. Not all business cases will or should include a feasibility study. Feasibility studies are usually conducted when the proposed project is highly complex, has a high potential for risk, or is a new type of project the organization has never undertaken before. Feasibility studies may be conducted as separate projects, as subprojects, or as a preproject phase. It's best to treat this activity as a project when the outcome is uncertain.

Identifying and Analyzing Stakeholders

Stakeholders are anyone who has a vested interest in the project. Stakeholders can include individuals as well as organizations, and both the project sponsor and the project manager are considered stakeholders. The project sponsor is the executive in the organization who authorizes the project to begin and is someone who has the ability and authority to assign funds and resources to the project. Identifying stakeholders is also a component of the project charter. Chapter 2 will go into much more detail about stakeholders, and Chapter 3 will talk about the project charter.

Project Selection

After the business case is created, you'll need some method to decide how you or the project selection committee will choose among competing projects.

Project selection methods are used to determine which proposed projects should receive approval and move forward. This process usually includes the allocation of high-level funding as well. Project selection may take place using formal documented guidelines, or it may be informal, requiring only the approval of a certain level of management.

Typically, a high-level board or committee will perform project selection. This committee may be cross-functional in nature and accountable for corporate-wide project selection, or selection may be determined on a departmental basis. A committee at the corporate level is composed of representatives from all departments, such as information technology, sales, marketing, finance, and customer service.

Project Selection Methods

A project selection committee uses a set of criteria to evaluate and select proposed projects. The selection method needs to be applied consistently across all projects to ensure the company is making the best decision in terms of strategic fit as well as the best use of limited resources.

Project selection methods will vary depending on the mission of the organization, the people serving on the selection committee, the criteria used, and the project itself. These methods could include examining factors such as market share, financial benefits, return on investment, customer satisfaction, and public perception. The exact criteria vary, but selection methods usually involve a combination of decision models and expert judgment.

Decision Models

A *decision model* is a formal method of project selection that helps managers make decisions regarding the use of limited budgets and human resources. Requests for projects can span a large spectrum of needs, and it can be difficult to determine a priority without a means of comparison. Is an online order entry application for the sales team more important than the addition of online help for the customer-support team? To the impacted departments, each project is probably viewed as a number-one priority. The problem is that there may not be adequate budget or staffing to complete both requests, and a decision must be made to approve one request and deny the other. Unless you can make an "apples-to-apples" comparison of the two requests, the decision will be very subjective. A decision model uses a fixed set of criteria agreed on by the project selection committee to evaluate the project requests. By using the same model to evaluate each project request, the selection committee has a common ground on which to compare the projects and make the most objective decision. You can use a variety of decision models, and they range from a basic ranking matrix to elaborate mathematical models.

There are two primary categories of decision models: benefit measurement methods and constrained-optimization models.

Benefit Measurement Methods

Benefit measurement methods provide a means to compare the benefits obtained from project requests by evaluating them using the same criteria. Benefit measurement methods are the most commonly used of the two categories of decision models. Four common benefit measurement methods are cost-benefit analysis, scoring model, payback period, and economic model.

Cost-Benefit Analysis   A *cost-benefit analysis* compares the cost to produce the product or service to the financial gain (or benefit) the organization stands to make as a result of executing the project. You should include development costs of the product or service, marketing costs, technology costs, and ongoing support, if applicable, when calculating total costs.

Let's say your proposed project involves developing and marketing a new product. The total costs are projected at $3 million. Based on market research, it appears the demand for this product will be high and that projected revenues will exceed the organization's goals. In this case, the cost-benefit analysis is positive and is a strong indicator you should select this project provided the business case justifies it as well.

The cost-benefit model is a good choice if the project selection decision is based on how quickly the project investment will be recouped from either decreased expenses or increased revenue. The weakness of using a cost-benefit analysis is that it does not account for other important factors, such as strategic value. The project that pays for itself in the shortest time is not necessarily the project that is most critical to the organization.

Scoring Model   A *scoring model* has a predefined list of criteria against which each project is rated. Each criterion is given both a scoring range and a weighting factor. The weighting factor accounts for the difference in importance of the various criteria. Scoring models can include financial data, as well as items such as market value, organizational expertise to complete the project, innovation, and fit with corporate culture. Scoring models have a combination of objective and subjective criteria. The final score for an individual project request is obtained by calculating the rating and weighting factor of each criteria. Some companies have a minimum standard for the scoring model. If this minimum standard is not obtained, the project will be eliminated from the selection process. A benefit of the scoring model is that you can place a heavier weight on a criterion that is of more importance. Using a high weighting factor for innovation may produce an outcome where a project with a two-year time frame to pay back the cost of the project may be selected over a project that will recoup all costs in six months. The weakness of a scoring model is that the ranking it produces is only as valuable as the criteria and weighting system the ranking is based on. Developing a good scoring model is a complex process that requires a lot of inter-departmental input at the executive level.

Payback Period   The *payback period* is a cash flow technique that identifies the length of time it takes for the organization to recover all the costs of producing the project. It compares the initial investment to the expected cash inflows over the life of the project and determines how many time periods elapse before the project pays for itself. Payback period is the least precise of all the cash flow techniques discussed in this section.

You can also use payback period for projects that don't have expected cash inflows. For example, you might install a new call-handling system that generates efficiencies in your call center operations by allowing the call center to grow over the next few years without having to add staff. The cost avoidance of hiring additional staff can be used in place of the expected cash inflows to calculate payback period.

Gustave Eiffel

The extraordinary engineer Gustave Eiffel put up the majority of the money required to build the Eiffel tower, nearly $2 million, himself. This was quite a sum in 1889, and his investment paid off. Tourism revenues exceeded the cost of constructing the tower in a little more than one year. That's a payback period any project manager would love to see. And Eiffel didn't stop there. He was wise enough to negotiate a contract for tourism revenues from the tower for the next 20 years.

Economic Model   An *economic model* is a series of financial calculations, also known as cash flow techniques, which provide data on the overall financials of the project. A whole book can be dedicated to financial evaluation, so here you'll get a brief overview of some of the common terms you may encounter when using an economic model: discounted cash flow, net present value, and internal rate of return.

Discounted Cash Flow   The *discounted cash flow* technique compares the value of the future worth of the project's expected cash flows to today's dollars. For example, if you expected your project to bring in $450,000 in year 1, $2.5 million in year 2, and $3.2 million in year 3, you'd calculate the present value of the revenues for each year and then add up all the years to determine a total value of the cash flows in today's dollars. Discounted cash flows for each project are then compared to other similar projects on the selection list. Typically, projects with the highest discounted cash flows are chosen over those with lower discounted cash flows.

Net Present Value   *Net present value (NPV)* is a cash flow technique that calculates the revenues or cash flows the organization expects to receive over the life of the project in today's dollars. For example, let's say your project is expected to generate revenues over the next five years. The revenues you receive in years 2, 3, and so on, are worth less than the revenues you receive today. NPV is a mathematical formula that allows you to determine the value of the investment for each period in today's dollars. Each period's resulting sum in present-day dollars is added together, and that sum is then subtracted from the initial investment to come up with an overall value for the project. The rule for NPV is that if NPV is greater than zero, you should accept the project. If it's less than zero, you should reject the project.

 The difference between NPV and discounted cash flows is that NPV subtracts the total cash flow in today's dollars from the initial project investment. Discounted cash flow totals the value of each period's expected cash flow to come up with a total value for the project in today's terms.

Internal Rate of Return   *Internal rate of return (IRR)* is the discount rate when the present value of the cash inflows equals the original investment. IRR states the profitability of an investment as an average percent over the life of the investment. The general rule is that projects with higher IRR values are considered better than projects with lower IRR values.

Constrained Optimization Models

Constrained optimization models are mathematical models, some of which are very complicated. They are typically used in very complex projects and require a detailed understanding of statistics and other mathematical concepts. A discussion of these models is beyond the scope of this book.

Expert Judgment

Expert judgment relies on the expertise of stakeholders, subject-matter experts, or those who have previous experience to help reach a decision regarding project selection. Typically, expert judgment is used in conjunction with one of the decision models discussed previously.

Companies with an informal project selection process may use only expert judgment to make project selection decisions. Although using only expert judgment can simplify the project selection process, there are dangers in relying on this single technique. It is not likely that the project selection committee members will all be authorities on each of the proposed projects. Without access to comparative data, a project approval decision may be made based solely on who has the best slide presentation or who is the best salesperson.

Political influence can also be part of the expert judgment. An executive with a great deal of influence may convince the selection committee to approve a particular project.

Once your selection committee has selected and approved a list of projects, the project manager will move forward with the project management processes. You'll look at one way these processes are organized in the next section.

Understanding the Project Management Knowledge Areas

The project manager is the person who oversees all the work required to complete the project by using a variety of tools and techniques. The *PMBOK Guide®* categorizes these tools and techniques into processes. For example, the Develop Project Charter process outlines inputs, tools and techniques, and outputs for producing the fully documented project charter, which is an output of this process.

The *Project Management Knowledge Areas* are collections of individual processes that have elements in common. For example, the Project Human Resource Management area is comprised of four processes that are all used to help establish, acquire, and manage project resources.

According to the *PMBOK Guide®*, these are the 10 Project Management Knowledge Areas:

- Project Integration Management
- Project Scope Management
- Project Time Management
- Project Cost Management
- Project Quality Management
- Project Human Resource Management
- Project Communications Management
- Project Risk Management
- Project Procurement Management
- Project Stakeholder Management

As you move through subsequent chapters, you will examine these areas in more detail. Keep in mind that these Knowledge Areas may not have equal importance on your next project. For example, if you are performing a project with internal resources and require a minimum amount of goods or supplies for the project, the Project Procurement Management area will have less emphasis in your project-planning activities than the Project Scope Management area.

Understanding the Role of the Project Manager

As stated earlier in this chapter, the project manager is the person responsible for integrating all the components and artifacts of a project and applying the various tools and techniques of project management to bring about a successful conclusion to the project. The project manager's role is diverse and includes activities such as managing the team; managing communication, scope, risk budget, and time; managing quality assurance; planning; negotiating; solving problems; and more.

Good soft skills are as critical to the success of a project as good technical skills. You'll examine many of the technical skills needed as they relate to the project management processes in the coming chapters, but I won't neglect to talk about the soft skills as well. These are skills any good manager uses on a daily basis to manage resources, solve problems, meet goals, and more.

You probably already use some of these skills in your day-to-day work activities. Here's a partial list:

- Leadership
- Communicating
- Listening
- Organization
- Time management
- Planning
- Problem-solving
- Consensus building
- Resolving conflict
- Negotiating
- Team building

Let's examine a few of these skills in a little more detail.

Leadership

A project manager must also be a good leader. Leaders understand how to rally people around a vision and motivate them to achieve amazing results. They set strategic goals, establish direction, and inspire and motivate others. Strong leaders also know how to align and encourage diverse groups of people with varying backgrounds and experience to work together to accomplish the goals of the project.

Leaders possess a passion for their work and for life. They are persistent and diligent in attaining their goals. And they aren't shy about using opportunities that present themselves to better their team members, to better the project results, or to accomplish the organization's mission. Leaders are found at all levels of the organization and aren't necessarily synonymous with people in executive positions. We've known our share of executive staff members who couldn't lead a team down the hall, let alone through the complex maze of project management practices. It's great for you to possess all the technical skills you can acquire as a project manager. But it's even better if you are also a strong leader who others trust and are willing to follow.

Communication

Most project managers will tell you they spend the majority of their day communicating. PMI® suggests that project managers should spend up to 90 percent of their time in the act of communicating. It is by far the number-one key to project success. Even the most detailed project plan can fail without adequate communication. And of all the communication skills in your tool bag, listening is the most important. Ideally, you've finely honed your leadership skills and have gained the trust of your team members. When they trust you, they'll tell you things they wouldn't have otherwise. As the project manager, you want to know everything that has the potential to affect the outcomes you're striving for or anything that may impact your team members.

Project managers must develop a communication strategy for the project that includes the following critical components:

- What you want to communicate
- How often you'll communicate
- The audience receiving the communication
- The medium used for communicating
- Monitoring the outcome of the communication

Keeping these components in mind and developing a comprehensive communication plan early in the project will help prevent misunderstanding and conflict as the project progresses.

We'll discuss communication in more detail in Chapter 8, "Communication Techniques."

Problem-Solving

There is no such thing as a project that doesn't have problems. Projects always have problems. Some are just more serious than others.

Early recognition of the warning signs of trouble will simplify the process of successfully resolving problems with minimal impact. Many times, warning signs come about during communications with your stakeholders, team members, vendors, and others. Pay close attention not only to what your team members are saying but also to how they're saying it. Body language plays a bigger part in communication than words do. Learn to read the real meaning behind what your team member is saying and when to ask clarifying questions to get the heart of the issue on the table.

We'll discuss specific techniques you can use to help with problem-solving and conflict resolution in Chapter 6, "Resource Planning and Management."

Negotiating

Negotiation is the process of obtaining mutually acceptable agreements with individuals or groups. Like communication and problem-solving skills, this skill is used throughout the life of the project.

Depending on the type of organizational structure you work in, you may start the project by negotiating with functional managers for resources. If you will be procuring goods or services from an outside vendor, you will likely be involved in negotiating a contract or other form of procurement document. Project team members may negotiate specific job assignments. Project stakeholders may change the project objectives, which drives negotiations regarding the schedule, the budget, or both. As you execute the project, change requests often involve complex negotiations as various stakeholders propose conflicting requests. There is no lack of opportunity for you to use negotiating skills during the life of a project, and you'll be learning about many of these examples in more detail in the coming chapters.

 Real World Scenario

Negotiating with the Business Unit

You're working on a project for the human resources department in your company. They'd like to streamline the recruitment process and set up a website for applicants to view the job descriptions and apply online. The hiring managers also need a streamlined way to quickly review resumes and applications and arrange for interviews with qualified candidates. You've gotten past the initial project request steps, and you're now in the process of determining the details of the requirements for the project.

You set up a meeting with the director of human resources. At the meeting you ask her two things. First, you want to know whether you can use someone from the business unit to assist you in understanding the business process flows. You make it clear that the assigned individual must be a subject-matter expert (SME) in the business process. Second, you ask whether you can have this individual full-time for one week. You suggest the name of someone whom you think will perform well as a business SME.

The director is surprised that you require so much time from one of her people. She asks you to more thoroughly explain your needs. You explain to her that in order for you to create a website that fully meets the business needs, you must understand how the business process works today and how it can be improved.

After some discussion back and forth, the two of you come to an agreement that you can have three days of someone's time and that you'll use two different business SMEs, splitting their efforts accordingly so that neither one has to fully dedicate their time to the business flow discovery process. The director stresses to you that her people are busy, and she is being generous in letting you have them at all.

You agree, thank her for her time, and get to work figuring out the best questions to ask the SMEs in order to complete the business flow discovery process in as efficient and timely a manner as possible.

Organization and Time Management

As stated earlier in this chapter, the project manager oversees all aspects of the work involved in meeting the project goals. The ongoing responsibilities of a typical project manager include tracking schedules and budgets and providing updates on their status, conducting regular team meetings, reviewing team member reports, tracking vendor progress, communicating with stakeholders, meeting individually with team members, preparing formal presentations, managing change requests, and much more. This requires excellent organizational skills and the ability to manage your time effectively. I've found that most project managers are good time managers as well, but if you struggle in this area, I strongly recommend taking a class or two on this topic.

Meetings consume valuable project time, so make certain they are necessary and effective. Effective meetings don't just happen—they result from good planning. Whether you conduct a formal team meeting or an individual session, you should define the purpose of the meeting and develop an agenda of the topics to be discussed or covered. It's good practice to make certain each agenda item has a time limit in order to keep the meeting moving and to finish on time. In my experience, the only thing worse than team members coming late to a scheduled meeting is a meeting that goes past its allotted time frame.

Clear documentation is critical to project success, and you'll want a system that allows you to put your hands on these documents at a moment's notice. Technology comes to the rescue with this task. Microsoft SharePoint is an excellent tool to help you organize project documents. Other tools are available to you as well. Find one that works, even if it's a manual system, and keep it up-to-date.

Summary

A project is a temporary endeavor that produces a unique product service or result. It has definitive start and finish dates. Project management is the application of tools and techniques to organize the project activities to successfully meet the project goals. A project manager is responsible for project integration and applying the tools and techniques of project management to bring about a successful conclusion to the project.

Organizational structures impact how projects are managed and staffed. The primary structures are functional, matrix, and projectized. The traditional departmental hierarchy in a functional organization provides the project manager with the least authority. The other end of the spectrum is the project-based organization, where resources are organized around projects; in these types of organizations, the project manager has the greatest level of authority to take action and make decisions regarding the project. The matrix organization is a middle ground between the functional organization and the project-based organization.

Programs are a collection or group of related projects that are managed together using coordinated processes and techniques. The collective management of a group of projects can bring about benefits that wouldn't be achievable if the projects were managed separately.

Portfolios are collections of programs, subportfolios, and projects that support strategic business goals or objectives. Portfolios may consist of projects that are not related.

Project selection techniques involve the use of decision models, such as a cost-benefit analysis and expert judgment, to allocate limited resources to the most critical projects.

Project managers are individuals charged with overseeing every aspect of a given project from start to finish. A project manager needs not only technical knowledge of the product or service being produced by the project but also a wide range of general management skills. Key general management skills include leadership, communication, problem-solving, negotiation, organization, and time management.

Exam Essentials

Be able to define a project.   A project brings about a unique product, service, or result and has definite beginning and ending dates.

Be able to identify the difference between a project and ongoing operations.   A project is a temporary endeavor to create a unique product or service. Operational work is ongoing and repetitive.

Be able to define a program and a portfolio.   A program is a group of related projects managed to gain benefits that couldn't be realized if they were managed independently. Portfolios are collections of programs, subportfolios, and projects that support strategic business goals or objectives. Programs and projects within the portfolio may not be related to one another.

Name the three types of organizational structures.   The three types of organizational structures are functional, matrix, and projectized structures. Matrix organizations may be structured as a strong matrix, weak matrix, or balanced matrix organization.

Be able to define the role of a project manager.   A project manager's core function is project integration. A project manager leads the project team and oversees all the work required to complete the project goals to the satisfaction of the stakeholders.

Be able to identify the most common project selection methods.   The most common project selection methods are benefit measurement methods such as cost-benefit analysis, scoring models, payback period, and economic models (which include discounted cash flows, NPV, and IRR), as well as expert judgment.

Understand what skills are needed to manage a project beyond technical knowledge of the product.   Key general management skills include leadership, communication, problem-solving, negotiation, organization, and time management.

Key Terms

Before you take the exam, be certain you are familiar with the following terms:

A Guide to the Project Management Body of Knowledge (PMBOK Guide®)

benefit measurement methods

colocated

constrained optimization models

cost-benefit analysis

decision model

discounted cash flow

economic model

expert judgment

feasibility study

functional organization

internal rate of return (IRR)

matrix organization

net present value (NPV)

operations

payback period

portfolio

program

project

project management

Project Management Institute (PMI®)

Project Management Knowledge Areas

project managers

project selection methods

projectized organization

scoring model

Review Questions

1. What is the definition of a project? Choose two.

   A. A group of interrelated activities that create a unique benefit to the organization

   B. Through the use of project management techniques, which are repeatable processes, a series of actions that are performed to produce the same result multiple times

   C. A temporary endeavor undertaken to create a unique product, service, or result

   D. A process used to generate profit, improve market share, or adhere to legal requirements

   E. A time-constrained endeavor with assigned resources responsible for meeting the goals of the project according to the quality standards

2. What organization is recognized worldwide for setting project management standards?

   A. PMC

   B. PMI®

   C. PMP

   D. CompTIA

3. What is the term for a group of related projects managed in a coordinated fashion?

   A. Life cycle

   B. Phase

   C. Process group

   D. Program

4. Which of the following are true regarding project portfolios? Choose two.

   A. The independent projects in the portfolio may not have anything in common.

   B. The programs in the portfolio are related to one another.

   C. The programs and projects within the portfolio support the strategic goals of the portfolio.

   D. An organization has only one portfolio.

   E. Portfolios consist of programs and do not contain stand-alone projects.

5. Which of the following general management skills does a project manager employ up to 90 percent of their time?

   A. Programming

   B. Communicating

   C. Leading

   D. Problem-solving

6. You receive a request from customer service to develop and implement a desktop management system for the customer-support staff. What type of project request is this?

   A. Business need

   B. Market demand

   C. Legal requirement

   D. Technological advance

7. You are working in a matrix organization. Choose two responses that describe this type of structure.

    A. Project resources are members of another business unit and may or may not be able to help you full-time.

    B. Matrix organizations can be structured as strong, weak, or balanced.

    C. Project managers have the majority of power in this type of structure.

    D. This organizational structure is similar to a functional organization.

    E. Employees are assigned project tasks by their project manager in this type of structure.

8. A project manager has the most authority under which organizational structure?

    A. Projectized

    B. Functional

    C. Balanced matrix

    D. Strong matrix

9. Your project has expected cash inflows of $7.8 million in today's dollars. Which cash flow technique was used to determine this?

    A. Discounted cash flow

    B. IRR

    C. NPV

    D. Cost-benefit analysis

10. Which of the following are the steps required to validate a project? Choose two.

    A. Analyze the feasibility.

    B. Justify the project.

    C. Align it to the strategic plan.

    D. Validate the business case.

    E. Identify and analyze stakeholders.

11. This general management skill concerns obtaining mutually acceptable agreements with individuals or groups.

    A. Leadership

    B. Problem-solving

    C. Negotiating

    D. Communicating

12. Federico, the director of the marketing department, has approached you with an idea for a project. What are the elements you'll include in the business case? Choose four.

    A. The business justification for the project

    B. The strategic opportunity/business need that brought about the project

    C. The recommended alternative

    D. List of key stakeholders

    E. Alternative solutions analysis

13. Your project has expected cash inflows of $1.2 million in year 1, $2.4 million in year 2, and $4.6 million in year 3. The project pays for itself in 23 months. Which cash flow technique was used to determine this?

   A. IRR

   B. NPV

   C. Discounted cash flow

   D. Payback period

14. You've been given an idea for a project by an executive in your organization. After writing the business-case analysis, you submit it to the executive for review. After reading the business case, he determines that the project poses a significant amount of risk to the organization. What do you recommend next?

   A. Proceed to the project selection committee.

   B. Reject the project based on the analysis.

   C. Proceed to writing the project plan.

   D. Perform a feasibility study.

15. You're a project manager working on a software development project. You are working hand in hand with a systems analyst who is considered an expert in her field. She has years of experience working for the organization and understands not only systems development but also the business area the system will support. Which person should make the decisions about the management of the project?

   A. Project manager

   B. Systems analyst

   C. Project manager with input from systems analyst

   D. Systems analyst with input from project manager

16. What is one disadvantage of a projectized organization?

   A. The organization doesn't work on anything that isn't project-related.

   B. Costs are high because specialized skills are required to complete projects in this type of structure.

   C. The functional managers have control over which team members are assigned to projects.

   D. Once the project is completed, the project team members may not have other projects to work on.

17. Which of the following are reasons for bringing about a project? Choose three.

   A. Feasibility study

   B. Market demand

   C. Business case justification

   D. Strategic opportunity

   E. Stakeholder needs

   F. Social needs

18. Your project has expected cash inflows of $7.8 million in today's dollars. The project's initial investment is $9.2 million. Which of the following is true?

    A. The discounted cash flows are lower than the initial investment, so this project should be rejected.

    B. The discounted cash flows are lower than the initial investment, so this project should be accepted.

    C. NPV is less than zero, so this project should be rejected.

    D. NPV is greater than zero, so this project should be accepted.

19. The executives in your organization typically choose which projects to perform first by reviewing the business case and then determining, based on their experience with similar projects, which will likely perform well and which will not. What form of project selection method is this?

    A. Business case analysis

    B. Expert judgment

    C. Feasibility analysis

    D. Decision model technique

20. Which two elements should always be included in a business case analysis? Choose two.

    A. Feasibility study

    B. Project selection methodology

    C. Alignment to the strategic plan

    D. Justification

    E. Cash flow techniques to determine financial viability

Chapter

2

Project Team Roles and Responsibilities

THE COMPTIA PROJECT+ EXAM TOPICS COVERED IN THIS CHAPTER INCLUDE

✓ **1.2 Classify project roles and responsibilities.**

- Sponsor/champion
  - Approval authority
    - Funding
    - Project charter
    - Baseline
    - High-level requirements
  - Control
  - Marketing
  - Roadblocks
  - Business case/justification
- Project manager
  - Manage team, communication, scope, risk, budget, and time
  - Manage quality assurance
  - Responsible for artifacts
- Project coordinator
  - Support project manager
  - Cross-functional coordination
  - Documentation/administrative support
  - Time/resource scheduling
  - Check for quality

- Stakeholder
  - Vested interest
  - Provide input and requirements
  - Project steering
  - Expertise
- Scheduler
  - Develop and maintain project schedule
  - Communicate timeline and changes
  - Reporting schedule performance
  - Solicit task status from resources
- Project team
  - Contribute expertise to the project
  - Contribute deliverables according to schedule
  - Estimation of task duration
  - Estimation of costs and dependencies
- Project Management Office (PMO)
  - Sets standards and practices for organization
  - Sets deliverables
  - Provides governance
  - Key performance indicators and parameters
  - Provides tools
  - Outlines consequences of nonperformance
  - Standard documentation/templates
  - Coordinate resources between projects

Stakeholder Roles and Responsibilities

Throughout the life of your project, you will interact with an important group of people: your stakeholders. Gaining stakeholder buy-in is critical for the success of your project. Let's talk a bit about who they are, what they want, and your role as the project manager in engaging with them.

> Remember from Chapter 1 that identifying and analyzing stakeholders is one of the steps in validating a project, along with preparing a business case.

A *stakeholder* is a person or an organization that has a vested interest in your project. In other words, they have something to gain or lose as a result of performing the project. They typically have a lot of influence. As you would expect, most of your stakeholders are concerned about the needs of their own departments (or organizations) first. They'll be looking to you as the project manager to help them understand how they'll benefit as a result of this project. If you are successful at winning the confidence and support of the project's key stakeholders, it will go a long way toward assuring the success of the project overall.

The general management skills discussed in Chapter 1 will come in handy when dealing with your stakeholders, particularly your communication and negotiation skills. Individual stakeholders may have different priorities regarding your project, and you may have to do some negotiating with your stakeholder groups to bring them to a consensus regarding the end goal of the project. Building consensus among a group with diverse viewpoints starts with up-front negotiation during the initial phases of the project and continues with ongoing communication throughout the life of the project. In Chapter 8 you will learn about communications in detail, including how you define and implement a communications plan geared to the needs of individual stakeholders.

Stakeholders are the people and organizations you will work with to determine project requirements. The expertise they bring from their respective business areas helps the project manager and project team when defining requirements, reviewing deliverables, and assuring the end product or result meets quality standards. Stakeholders provide direction throughout the life of the project and review and approve the final end product, service, or result.

The Customer

Your customer is also a stakeholder. The *customer* is the recipient of the product or service created by the project. In some organizations this stakeholder may also be referred to as the *client*. A customer is often a group or an organization rather than a single person. Customers can be internal or external to the organization.

Unfortunately, some stakeholders may not support your project, for any number of reasons. They may not like the person who requested the project, they may not like the goal of the project, it might create major change in their business unit, it might change their operational procedures, and so on. A project that creates a major impact on operational procedures may be viewed as a threat. In fact, any project that brings about a major change in the organization can cause fear and generate resistance.

The key is to get to know your stakeholders as soon as possible. Set up individual meetings or interviews early on to understand their perspectives and concerns about the project. Their concerns aren't going to go away, and if you ignore some of your stakeholders, the issues they raise will become more and more difficult to resolve as the project progresses. Take the time to meet with them regularly. This will help you to set and clarify expectations and help them see the benefits of moving forward with the project.

The Project Sponsor

The project *sponsor* is usually an executive in the organization who has the authority to assign money and resources to the project. The sponsor may also serve as a *champion* for the project within the organization. The sponsor is an advisor to the project manager and acts as the tiebreaker decision maker when consensus can't be reached among the stakeholders. One of the primary duties of a project manager is keeping the project sponsor informed of current project status, including conflicts or potential risks.

A project champion is usually the project sponsor or one of your key stakeholders. They spread the great news about the benefits of the project and act as a cheerleader of sorts, generating enthusiasm and support for the project.

The project sponsor/champion typically has the following responsibilities:

- Provide or obtain financial resources
- Approve the project charter
- Approve the project baseline
- Help define and approve the high-level requirements
- Define the business case and justification for the project
- Authorize assignment of human resources to the project
- Assign the project manager and describe their level of authority
- Serve as final decision maker for all project issues

- Negotiate support from key stakeholders
- Communicate or market the benefits of the project
- Monitor and control delivery of major milestones
- Run interference and remove roadblocks

> The *project baseline* includes the approved schedule, cost, scope, and quality plans and documents. The project baseline is then used to measure performance as the project progresses. You can refer to the project baseline at any time to determine whether you are on schedule, within scope, within budget, and determine whether the quality standards are on target.

The Project Manager

Chapter 1 talked about the project manager in detail. You will recall that this is the person responsible for managing the work associated with the project. The following is a brief list of the project manager's responsibilities and is not all-inclusive. Each of these items will be discussed in more detail in the remaining chapters of this book.

Managing the Project Team   The *project team* consists of members from inside and sometimes outside the organization. You will not always have the ability to choose your project team. Remember that in a functional organization, the most common type of organizational structure, other managers will assign resources to your team.

Communicating with Stakeholders and Project Team Members   Keeping the sponsor, stakeholders, and team members up to date on current project status, issues, and other information is a key responsibility of the project manager. Communicating starts the moment a project idea is formulated and continues through to final close-out and approval of the project.

Managing Scope   Scope describes the goals, deliverables, and requirements of the project. The project manager will capture and document scope, along with help from the stakeholders and team members, and will monitor and manage project scope using change-control processes.

Managing Risk   Risks are events that may occur that would impact the project, either positively or negatively, and generally have consequences if they occur. Risks are identified, managed, tracked, and monitored by the project manager.

Managing the Project Budget   Project managers are responsible for monitoring and tracking project costs and alerting the sponsor as soon as possible if costs are higher than expected or the project is running through funds faster than anticipated.

Managing the Schedule   Most project sponsors and executive-level stakeholders love the saying "on time and on budget" and will often chant this phrase in project meetings. Along with managing the budget, you are also responsible for managing the schedule and assuring that key deliverables are performed on time.

Managing Quality Assurance   It isn't enough to deliver on time and on budget. You will also be responsible for assuring the quality of the deliverables and making certain they meet quality standards and are fit for use.

Project Artifacts   Project *artifacts* are the documents, templates, agendas, diagrams, and other work products used in managing the project. For example, the project charter is an artifact. The project manager is responsible for maintaining and archiving the artifacts of the project.

The project manager is responsible for many aspects of the project. The most important include making sure stakeholders are satisfied with the deliverables and end product, service, or result of the project; integrating the work of the project; and communicating with stakeholders.

 It's your responsibility as the project manager to manage stakeholder expectations throughout the project.

The Project Coordinator

The *project coordinator* assists the project manager in all aspects of the project. They generally assist the project manager in an administrative support function. For example, they may coordinate cross-functional team members and resources, set up meetings, help the project manager with resource scheduling, monitor the schedule, and check for quality products and processes.

A project coordinator position is generally a first step for those looking to pursue a career in project management. It's a great way to get a close up look at project management.

The Project Scheduler

The project *scheduler* is responsible for developing and maintaining the project schedule, communicating the timelines and due dates, reporting schedule performance, and communicating any schedule changes to the stakeholders and project team members. Project schedulers obtain status updates in terms of timelines and due dates from the team members assigned to the work of the project so they can keep the schedule updated.

Typically, project schedulers are employed in large organizations or on very large projects. They may also work for the project management office (we'll discuss this shortly) as a shared resource who works with all the project managers assigned to the project management office.

The Project Team Members

Project team members are the experts who will be performing the work associated with the project. Depending on your organizational structure, these people may report directly to the project manager, report to a functional manager within the organization, or work in a matrix-managed team.

If your resources are supplied by another part of the organization, the functional managers who assign those resources are critical stakeholders. You need to establish a good relationship with your functional managers and brush up on those negotiation skills because you'll need them. Normally, more than one project manager is competing for the same resource pool. So, it's a good idea to document your agreements with the functional manager regarding the amount of time the resource will be available for your project, as well as the deliverables they're accountable for, in order to prevent future misunderstandings. You should also obtain prior agreement regarding your input to the employee's annual performance appraisal, salary increase, and bonus opportunity.

Project team members may be assigned to the project either full-time or part-time. Most projects have a combination of dedicated and part-time resources. If you have part-time resources, you need to understand their obligations outside the project and make certain they are not over-allocated.

Project team members are responsible for a number of activities on the project. One of their most important duties is contributing deliverables according to the schedule. A partial list of other duties may include the following:

- Time and duration estimates for the tasks they are working on

- Cost estimates for deliverables or other project work purchased from outside the organization

- Status updates on the progress of their tasks

- Dependencies related to their tasks

The Project Management Office

Many organizations have a *project management office (PMO)* in place that manages projects and programs. The PMO provides guidance to project managers and helps present a consistent, reliable approach to managing projects across the organization. PMOs are responsible for maintaining standards, processes, procedures, and templates related to the management of projects. They are responsible for identifying the various projects across the organization and including them within a program, where appropriate, to capitalize on the collective benefits of all the projects within the program.

Some of the functions a PMO may provide are as follows:

- Project management standards and processes

- Tools, templates, and artifacts to help manage projects consistently

- Setting deliverables

- Governance process for managing projects and setting priorities

- Key performance indicators and metrics

- Standards of performance including consequences of nonperformance

- Coordination of resources among projects

A complete list of stakeholders varies by project and by organization. The larger and more complex your project is, the more stakeholders you will have. Sometimes you will have far more stakeholders than you want or need, especially on high-profile projects. We recommend you define who you think the stakeholders are on the project and review the list with your project sponsor. The project sponsor is often in a better position to identify those stakeholders who are influential people in the organization. These types of stakeholders can make or break your project, and your sponsor can assist you in identifying their needs.

As you can see, your project team members and stakeholders represent a wide range of functional areas and a diverse set of wants and needs relative to your project. To keep track of everyone, you may want to develop a stakeholder matrix. You'll look at that next.

The Stakeholder Matrix

If you have a large project with multiple stakeholders, it may be appropriate to create a stakeholder matrix to help you keep track of everyone. You can use a simple spreadsheet to create the matrix. At a minimum, it should include a list of all the project stakeholders with the following information for each one:

- Name
- Department
- Contact information
- Role on the project
- Needs, concerns, and interests regarding the project
- Level of involvement on the project
- Level of influence over the project
- Notes for your own reference about future interactions with this stakeholder, political issues to be aware of, or individual quirks you want to remember about this stakeholder

Since project stakeholders can move on and off the project at different times, it's important that the project manager reviews and updates the matrix periodically.

 Real World Scenario

The Enterprise Resource Planning Implementation

Your organization is considering implementing an enterprise resource planning (ERP) system. This system will handle all the back-office functions for your organization, including procurement, human resources, materials inventory, fleet management, budgeting, and accounting. Currently, your organization has 14 disparate systems that handle these functions.

Identifying your stakeholders for this project turns out to be a daunting task. Within each of the departments, there are executives with their own ideas regarding the system requirements, and there are also functional managers who are much closer to their processes and business unit functions on a day-in and day-out basis. For the most part, their requirements match those of the executives, but you are having some difficulty reconciling the day-to-day processing requirements with some functionality the executives have requested.

Project success or failure can rest on any one of these stakeholders. As the project manager, your best course of action is to meet individually with each stakeholder and understand their individual requirements and concerns about the project. Next, you'll document those requirements and concerns and then bring the key stakeholders together to discuss where they agree and gain consensus regarding the differences.

Receiving a Project Request

Most projects start with an idea. The idea generates a project request, which typically starts with a business case, and the process to review and authorize the project begins. You've already learned about identifying the project, developing the business case, validating the business case, and using project selection techniques in Chapter 1. Now you'll learn about the project request process and documenting the high-level scope definition and the high-level project requirements.

The Project Request Process

The project request process can be formal or informal, depending on the organization. You may have a process that requires a formal written document describing the project goals and justification, or you may experience what I like to call *drive-bys*—the project your boss tells you about in the 30-second elevator ride to the lobby. Regardless of who initiates a project request or how it's initiated, the organization must review it and make a decision on a course of action. To do that, you'll need to gather enough information to adequately evaluate the request and determine whether the project is worth pursuing.

After receiving a request, your next step involves meeting with the requestor to clarify and further define the project needs, identify the functional and technical requirements, and document the high-level requirements.

 Several needs or demands bring about the reasons for a project. Refer to Chapter 1, "Initiating the Project," if you need a refresher.

The High-Level Scope Definition

The high-level scope definition describes the project objectives, high-level deliverables, and the reason for the project.

This description explains the major characteristics of the product or service of the project and describes the relationship between the business need and the product or service requested. Before you jump into completing your high-level deliverables and requirements, you need to make certain the problem or need that generated the project request is clearly defined and understood. That means you'll need to meet with the person requesting the project to clarify the project goals and understand what problem they're trying to solve. If the problem is unclear, the solution may be off target, so it's critical that you understand the problem before you move on to defining requirements.

Defining the Problem

A project can get off to a bad start if the project manager does not take the time to clearly define the problem or need that has generated the project request. Have you ever been on a project where people are working furiously to meet a deadline but no one appears to know why the work is being done? Then halfway through the project everything changes or, worse yet, the whole thing gets canceled? If this sounds familiar, it may be that a solution was being developed without clarifying the problem.

The customer is not always able to articulate what the problem is, and their request may be vague and loosely worded. Your job as project manager is to figure out what the customer really means. I've worked on many IT-related projects that were developed and implemented to the precise requirements of the customer only to hear them exclaim, "You delivered what I asked for but not what I wanted!"

It's your responsibility to investigate the customer request and communicate your understanding of the request. This may result in the creation of a project concept document, or perhaps the business case, which represents your first attempt at restating the customer request to demonstrate understanding of the project.

Problems can also arise when project requests are proposed in the form of a solution. It is not uncommon for customers to come to you with a specific request and a solution they've already discovered that satisfies the request. You may be thinking this is great news because there is no need to tie up your calendar with a lot of requirements-gathering meetings. The problem is, your customer may not be asking for the right solution. As a project manager, you need to make certain that the problem has been identified before the solution is proposed.

Let's say you get a request for a new billing system. The first thing you should do is meet with the person making the request to get more information. Why do they need a new billing system? What functionality is missing from the existing system? What business need or opportunity do they believe this new system will solve? These kinds of questions will help you understand what is behind the new billing system request. If your project requestor is concerned about the number of customer calls related to general billing questions, the best solution might not be a new system but rather a clearer explanation of the charges. If they are interested in a new look and feel for the bill, you may be dealing with requirements that range from reformatting the current bill data to using a different paper to print bills. Numerous business needs may cause your customer to want a "new billing system," but many of them may have nothing to do with developing an entirely new application. That is why a good project manager asks questions to uncover what is behind a request. Lack of up-front clarification and problem definition has been the downfall of numerous projects. Do not assume that a customer-requested solution is always the best solution until you understand the business need.

Documenting the High-Level Scope Definition

Once you understand the answers to the questions posed in the previous section, you should have a good understanding of the goals and objectives of the project. You can document the reason for the project, state the problem you're trying to solve, and provide a high-level description of some of the deliverables needed to make the project a success. This information will serve as a basis for the project charter, which you'll learn about in the next chapter.

Next you'll begin a case study that will continue throughout the remainder of the book. The case study will outline the concepts you've learned in the chapter and walk you through a simulated project based in a fictitious business.

 Real World Scenario

Main Street Office Move

This case study will appear throughout the remainder of the book. It's designed to review the use of most of the project management elements covered, using information from each chapter you read.

Your organization is moving all their offices to a new, centralized location. Currently, there are staff members in three different buildings located throughout the city. You are the project manager and will coordinate, communicate, and manage all aspects of the project.

Your project sponsor is Kate Anderson. She is the CFO and wrote the business case for this project. Kate sent you a copy of the business case to review. You picked out a few key points on first reading: The move will be completed by December 31. There should be minimal disruption to employees. The cost benefit to the organization is significant because the new building is energy efficient and, more importantly, all employees are together so there is no need to travel between locations for meetings. This will save on fuel costs, reduce the size of the fleet, and increase productivity.

Your first step is to set up a meeting with Kate to discover her goals for this project, document a high-level scope definition, and understand any concerns she may have about this project. You will document this information as you know you will be able to use it for both the project request process and the project charter document. You will also ask Kate who she believes the key stakeholders are on this project and document them in your stake-holder matrix.

After your review of the business case, you have identified the following key stakehold-ers and will confirm this with Kate: the human resources manager, the marketing director, and the facilities director.

After Kate's meeting, you will meet with the key stakeholders to determine their goals and expectations for the project.

In the meantime, you begin preparing the project request and will submit this through the formal request process. The business case and project request are artifacts of the project, and you save these in the project repository.

Summary

Project stakeholders are anyone who has a vested interest in the outcomes of the project. Some project stakeholders you will likely encounter include the project sponsor, team mem-bers, functional managers, and customers (both internal and external).

A project sponsor is an executive in the organization who has the authority to assign budget and resources to the project. Project sponsors serve as the final decision-makers on the project, sign and approve the project charter, and remove obstacles so the team can perform their work. Project sponsors often act as the project champions as well. They spread enthusiasm for the project and act as cheerleaders regarding its benefits.

The project coordinator assists the project manager with administrative functions. The project scheduler coordinates and updates the project schedule and keeps stakeholders informed of schedule progress. The project manager is responsible for coordinating and managing the project team, communications, scope, risk, budget, and time. They also man-age quality assurance and are responsible for the project artifacts. A PMO provides guide-lines, templates, and processes for managing the project.

The project request process starts with a high-level scope definition that describes the project objectives, the high-level deliverables, and the reason for the project. It also describes the relationship between the business need and the product or service requested.

Exam Essentials

Be able to define a project manager. The project manager manages the team, communica-tion, scope, risk, budget, and time. They also manage quality assurance and are responsible for the project artifacts.

Be able to define a project sponsor. A project sponsor is an executive in the organization who has the authority to allocate dollars and resources to the project. The sponsor approves funding, the project charter, the project baseline, and high-level requirements. They have final decision-making authority for the project, help with marketing the benefits of the project, remove roadblocks for the team, and participate in business case justification.

Be able to define project stakeholders. A stakeholder is anyone who has a vested interest in the project and has something to gain or lose from the project. Stakeholders include the sponsor, project manager, project team members, functional managers, customers, team members, and others with an interest in the project.

Be able to define a project coordinator. Project coordinators assist the project manager with cross-functional coordination, documentation, administrative support, time and resource scheduling, and quality checks.

Be able to define a scheduler. The scheduler is responsible for developing and maintaining the project schedule, communicating timeline and changes, reporting on schedule performance, and obtaining the status of work performed from team members.

Be able to define the project team. The project team contributes expertise to the project, works on deliverables according to the schedule, estimates task durations, estimates costs, and estimates dependencies.

Be able to define the project management office. The PMO provides guidance to project managers and helps present a consistent, reliable approach to managing projects across the organization. PMOs are responsible for maintaining standards, processes, procedures, and templates.

Key Terms

Before you take the exam, be certain you are familiar with the following terms:

artifacts

champion

client

customer

project baseline

project coordinator

project management office (PMO)

project team

scheduler

sponsor

stakeholder

Review Questions

1. You have identified all the key stakeholders on the project. You've listed their names, departments, interests in the project, and level of influence. What have you created? Choose two.

    A. A project request

    B. A stakeholder matrix

    C. An artifact

    D. A business case

    E. A high-level scope definition

2. This person is responsible for authorizing the project to begin and for signing the project charter.

    A. Project sponsor

    B. Executive in the organization who requested the project

    C. Project champion

    D. Project manager

3. This team member assists the project manager with administrative functions.

    A. Project management office

    B. Project scheduler

    C. Project coordinator

    D. Project sponsor

4. A key responsibility of the project manager includes informing this person of changes, status, conflicts, and issues on the project.

    A. The project requestor

    B. The project sponsor

    C. The project champion

    D. The most influential project stakeholder

5. Which of the following does a PMO provide? Choose three.

    A. High-level scope definition

    B. Business case

    C. Tools

    D. Governance process

    E. Change control

    F. Templates

6. Which stakeholder assigns employees to the project?

   A. Project manager

   B. Functional manager

   C. Customer

   D. Sponsor

7. The project sponsor (who is also the project champion in this question) provides which of these functions to the project? Choose three.

   A. Preparing the stakeholder matrix

   B. Marketing the project

   C. Removing roadblocks

   D. Assigning project team members

   E. Defines the business justification

   F. Responsible for artifacts

8. The high-level scope definition describes which of the following?

   A. Reason for the project

   B. High-level deliverables of the project

   C. Objectives of the project

   D. All of the above

9. You are the project manager and have hired a team member to provide cross-functional coordination, check for quality, and assist you with documentation. What role does this person hold?

   A. Project scheduler

   B. PMO team member

   C. Project coordinator

   D. Project team member

10. You receive a confusing request from the marketing department to develop a new product campaign. What is the first step you should take?

   A. You meet with the marketing person to identify and clarify the request.

   B. You write the project charter.

   C. You submit a request to the project selection committee.

   D. You request the finance department to do a cost-benefit analysis.

11. Jack is a key stakeholder on the project. He's a big believer in the benefits of the project and provides enthusiasm, energy, communication, and motivation for your project. What is Jack's role?

   A. Project stakeholder

   B. Project champion

   C. Project team member

   D. Business analyst

12. You're working on a project to develop a client-server application. Your company's collections department will equip their field personnel with wireless tablet PCs that will connect with the server and database. You work for the PMO and are managed by the director of administrative services. The IT department is managed by the director of IT. The telecommunications segments are managed by the director of telecommunications. Who is the project sponsor?

    A. Manager of collections

    B. Director of administrative services

    C. Director of IT

    D. Director of telecommunications

    E. Not enough information

13. These people manage quality assurance, scope, risk, budget, and time.

    A. Project team members

    B. Project coordinators

    C. Project schedulers

    D. Project manager

14. This stakeholder is the recipient of the product or service created by the project.

    A. Sponsor

    B. Customer

    C. Client

    D. B and C

    E. All of the above

15. Project team members are responsible for several aspects of the project. Choose three.

    A. Contribute deliverables according to the schedule

    B. Set standards and practices

    C. Cross-functional coordination

    D. Provide input and requirements

    E. Estimate task durations

    F. Estimate costs

16. Rahul is a key resource for your project. You've worked with him on past projects and have identified him as one of the team members for this project. Some of Rahul's duties will include reporting on schedule performance, obtaining task status from team members, communicating timelines, and communicating changes to timelines. What is Rahul's role on this project?

    A. Project scheduler

    B. Project manager

    C. Project coordinator

    D. PMO team member

17. Your project team members and key stakeholders are confused about the project you are all working on. What is the most compelling reason for this?

   A. The problem or need generating the project request was not defined.

   B. The project sponsor has not signed the project request authorizing it to begin.

   C. The business case is not well documented.

   D. The project team members don't understand their roles and responsibilities.

18. The governance processes for project management are well established in your organization. Who established these processes?

   A. Project team

   B. Project manager

   C. Sponsor

   D. PMO

19. This component of the project request describes the project objectives, the high-level deliverables, and the reason for the project.

   A. Project request justification

   B. High-level deliverables definition

   C. High-level scope definition

   D. High-level goals and objectives definition

20. Which of the following are true regarding the project baseline? Choose three.

   A. The project baseline is approved by the project manager.

   B. The project baseline includes the approved project scope.

   C. The project baseline includes the approved business justification.

   D. The project baseline includes the approved project quality plan.

   E. The project baseline includes the approved project request.

   F. The project baseline includes the approved project schedule.

   G. The project baseline is approved by the project sponsor.

Chapter 3

Creating the Project Charter

THE COMPTIA PROJECT+ EXAM TOPICS COVERED IN THIS CHAPTER INCLUDE

✓ **1.3 Compare and contrast standard project phases.**

- Initiation
  - Project charter
  - Business case
  - High-level scope definition
  - High-level risks
- Planning
  - Schedule
  - Work breakdown structure
  - Resources
  - Detailed risks
  - Requirements
  - Communication plan
  - Procurement plan
  - Change management plan
  - Budget
- Execution
  - Deliverables
- Monitor and control
  - Risks/issues log
  - Performance measuring and reporting
  - Quality assurance/governance

- Change control
- Budget
- Closing
    - Transition/integration plan
        - Training
    - Project sign off
    - Archive project documents
    - Lessons learned
    - Release resources
    - Close contracts

✓ **4.2 Given a scenario, analyze project centric documentation.**

- Project charter
- Project management plan
- Issues log
- Organizational chart
- Scope statement
- Communication plan
- Project schedule
- Status report
- Dashboard information
- Action items
- Meeting agenda/meeting minutes

Project Process Groups

PMI® defines project management as a series of processes that are executed to apply knowledge, skills, tools, and techniques to the project activities to meet the project requirements. These processes have been organized into five process groups: Initiating, Planning, Executing, Monitoring and Controlling, and Closing. Each process group contains at least two individual processes (some have many more) that have their own inputs to the process, tools and techniques, and outputs. For example, the Develop Project Charter process is grouped within the Initiating process group. It has five inputs, two tools and techniques, and one output (the project charter).

Don't be confused by terms. PMI® calls these processes *process groups*, whereas CompTIA calls them *phases*. They have the same outputs no matter what they're called. I will use the terms *process groups* and *phases* interchangeably throughout the book.

The process groups or phases are tightly linked. Outputs from one group usually become inputs to another group. The groups may overlap, or you may find that you have to repeat a set of processes within a process group. For example, as you begin executing the work of the project, you may find that changes need to be made to the project management plan. So, you may repeat some of the processes found in the Planning process group and then reperform many of the Executing processes once the changes to the plan are made. This is known as an *iterative* approach. Figure 3.1 shows the links between the groups.

As you learn about each group, notice the correlation between the process groups (or phases) and the domains covered in the CompTIA Project+ exam. These phases are the foundation of project management. You need to understand each group, its characteristics, and how it contributes to delivering the final product, service, or result of the project.

FIGURE 3.1   PMI® process groups

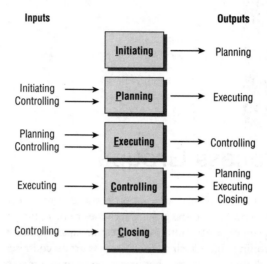

The Initiating Processes

Initiation is the formal authorization for a new project to begin or for an existing project to continue into the next phase. The Initiating phase is the first of the five process groups that PMI® describes in the *PMBOK Guide®*.

The *Initiating* processes include all the activities that lead up to the final authorization to begin the project, starting with the original project request. This process can be formal or informal, depending on the organization. The key activities in the Initiating process are as follows:

▪ Creating the project charter

▪ Creating a business case and justification

▪ Defining a high-level scope definition

▪ Identifying high-level risks

Chapter 1 already covered the business case. You'll learn more extensively about the project charter, high-level scope definition, and high-level risks throughout the remainder of this chapter.

Make certain you understand the key activities of each of the project phases for the exam.

The Planning Processes

In the *Planning* processes, the project goals, objectives, and deliverables are refined and broken down into manageable units of work. Project managers create time and cost estimates and determine resource requirements for each activity. Planning involves several other critical areas of project management, including communication, risk, human resources, quality, and procurement.

The key activities in the Planning process group are as follows:

- Develop a project schedule.

- Create a work breakdown structure.

- Determine resources.

- Identify and plan for detailed risks.

- Determine project requirements.

- Write a communication plan.

- Develop a procurement plan if utilizing resources outside the organization.

- Develop a change management plan.

- Define the project budget.

The Planning process group is unquestionably one of the most critical elements of managing a project. It's possible that a project manager will spend as much time planning the project as performing the work of the project (sometimes more).

The Planning phase contains many processes that all generally lead to the creation of plans or documents that are used throughout the project to ensure that goals of the project are being met. Table 3.1 shows the documents that CompTIA highlights in their objectives along with their purpose, and the process group where they are typically produced. You will learn about each of the project management documents shown here throughout the remainder of the book.

TABLE 3.1   Project management documents

Document Name	Document Description	Process Group
Project charter	Authorizes the project to begin	Initiating
Project management plan	Consists of all the project planning documents such as charter, scope statement, schedule, and more	Planning
Issues log	A list of issues, containing list numbers, descriptions, and owners	Executing
Organizational chart	Describes project team member organization and reporting structures	Planning

TABLE 3.1   Project management documents *(continued)*

Document Name	Document Description	Process Group
Scope statement	Documents the product description, key deliverables, success and acceptance criteria, key performance indicators, exclusions, assumptions, and constraints	Planning
Communication plan	Documents the types of information needs the stakeholders have, when the information should be distributed, and how the information will be delivered	Planning
Project schedule	Determines the start and finish dates for project activities and assigns resources to the activities	Planning
Status report	A report to stakeholders on the status of the project deliverables, schedule, risks, issues, and more	Executing
Dashboard information	An electronic reporting tool that lets users choose elements of the project to monitor project health and status	Executing
Action items	A list of project actions that should be resolved in order to fulfill deliverables	Executing
Meeting agenda/meeting minutes	Meeting agendas describe the items to be discussed and addressed at upcoming meetings, and minutes recap what was discussed and the decisions made at the meeting.	Executing

All the documents in this table will be discussed throughout the remainder of this book.

The Executing Processes

The *Executing* processes are where the work of the project is performed. The project manager must coordinate all the project team members as well as other resources assigned to the project.

Executing processes include the actual execution of the project management plan, team development, quality assurance, information distribution, and more. The key activities in the Executing process are as follows:

- Deliverables are produced and verified.

Deliverables are produced and verified during this process. If they do not conform to expectations, change requests are entered or corrective actions are taken (both of which occur in the Monitoring and Controlling processes) to ensure the deliverables adhere to specifications.

Resource management is important during the Executing processes. You'll acquire the project team members during this process, make certain they are utilized appropriately, and perform team-building activities. This process also includes working with vendors and contractors who are external to the organization.

The Monitoring and Controlling Processes

The *Monitoring and Controlling* processes are the activities that monitor the progress of the project to identify any variances from the project management plan. Requests for changes to the project scope are included in this process group. This area is also where corrective actions are taken to get the work of the project realigned to the project plan.

Other areas of the Monitoring and Controlling phase include scope control, cost control, schedule control, quality control, performance reporting, and risk control.

The key activities in the Monitoring and Controlling processes are as follows:

- Monitoring the risks/issues log
- Performance measuring and reporting
- Performing quality assurance/governance activities
- Administering the change control process
- Monitoring the budget

The Closing Processes

The primary purpose of the *Closing* processes is to document the formal acceptance of the project work and to hand off the completed product to the organization for ongoing maintenance and support.

Closing processes include project sign-off, archive of project documents, turnover to a maintenance group, release of project team members, and review of lessons learned.

The key activities in the Closing process are as follows:

- Transition/integration plan to the maintenance/operations team
- Training for those who will support the product, service, or result of the project once it's turned over to operations
- Project acceptance and sign-off

- Archiving project documents
- Documenting lessons learned
- Releasing resources
- Closing contract

This process group or phase is the one most often skipped in project management. And although some of these activities may seem fairly straightforward, several elements of this process group deserve close attention. Chapter 10, "Project Tools and Documents," will explore the last stages of a project.

Use the old (and now outdated) poison antidote, syrup of ipecac, to help you remember the process groups in order: Initiating, Planning, Executing, (Monitoring and) Controlling, and Closing.

Creating the Project Charter

The result of the Initiating process is the *project charter*. This document provides formal approval for the project to begin and authorizes the project manager to apply resources to the project. The project sponsor is the one who publishes, signs, and approves the project charter. Publishing the charter is a major milestone because it is the first official document of your approved project.

For the exam, remember that the project sponsor is the person who authorizes and approves the project charter. The project manager or the person who requested the project is typically the one who writes the project charter and makes certain it's distributed to all the key stakeholders, but the sponsor is the one who approves it.

Organizational standards may drive the specific format of the project charter and the information it contains. As a project manager, you should check with the PMO to determine whether there is a template or a required format for the project charter.

The following are the key elements that should be included in your project charter. Chapter 1 talked about the purpose or justification for the project. This is documented in the business case and can easily be copied into the project charter. The next section will cover the remaining elements.

- Purpose or justification for the project
- Project goals and objectives
- Project description
- Key deliverables

- High-level list of requirements
- High-level milestones
- High-level budget
- High-level assumptions
- High-level constraints
- High-level risks
- Name of the project manager and their authority level
- Name of the sponsor
- Criteria for project approval

Goals and Objectives

The charter documents the high-level goals and objectives of the project. A project charter needs to include a clear statement as to what end result the project will produce and how success will be measured. Goals and objectives must be clear and stated in such a manner that the end result is easily measured against the objective. Instead of stating "Build a new highway," the goal should include measurable outcomes like "Build a new highway between City A and City B that has three lanes in both directions by June 30."

Working with the sponsor to document quantifiable and measurable goals is key to the project success. It gives the customer, sponsor, key stakeholders, project manager, and team members the same common understanding of the end result of the project.

Project Description

The *project description* documents the key characteristics of the product, service, or result that will be created by the project. The project description also documents the relationship between the product being created and the business need that drove the project request. This description needs to contain enough detail to be the foundation for the Planning process group, which begins once the charter is signed.

 The project description in the charter starts out at a high level, and more details are added once you develop the project scope statement, which is discussed in Chapter 4.

Key Deliverables

Deliverables are measurable outcomes or results or are specific items that must be produced in order to consider the project complete. Deliverables are tangible and are easily measured and verified. For example, let's say your project involves manufacturing a new garden cart.

One of the components of the cart is wheels. Because of the design of your cart, the wheels must be 12 inches in diameter. This is a tangible, verifiable deliverable that must be met in order for the project to be a success.

Getting the deliverables and the requirements correct are critical to the success of your project. No matter how well you apply your project management skills, if the wrong deliverables are produced or the project is managed to the wrong objectives, you will have an unsuccessful project on your hands (and will probably need to update your resume).

High-Level Requirements

Requirements describe the characteristics of the deliverables that must be met in order to satisfy the needs of the project. Requirements might also describe results or outcomes that must be produced in order to satisfy a contract, specification, standard, or other project document (typically, the scope statement). Requirements quantify and prioritize the wants, needs, and expectations of the project sponsor and stakeholders.

The project charter contains a high-level look at the requirements. As you progress in the planning of the project, more information will become known, and the requirements will become much more detailed. I will talk more about requirements in Chapter 4, "Creating the Work Breakdown Structure."

High-Level Milestones

Milestones are major events in a project that are used to measure progress. They may also mark when key deliverables are completed and approved. Milestones are also used as checkpoints during the project to determine whether the project is on schedule.

High-Level Budget

The detailed project budget is prepared later during the Planning processes. But for the purposes of the project charter, you need to have a high-level estimate of the project's costs. You can use historical information from past projects that are similar in size, scope, and complexity to the current project. Or you may ask your vendor community to help you with some high-level estimates for the project.

High-Level Assumptions

Assumptions are events, actions, concepts, or ideas you believe to be true and plan for. For example, you may have a resource need for the project with a highly specialized skill. Someone with this skill set resides in your maintenance department, and since you've worked with both the functional manager and this resource on past projects, you assume they'll be available for this project. You can make assumptions about many elements of the project, including resource availability, funding, weather, timing of other related events, availability of vendors, and so on. It's important to always document and validate your project assumptions.

🌐 **Real World Scenario**

Planning a School Building Repair

You have been assigned a project that requires repairs to the roof of a school building and replacement of the air conditioning and airflow cleaning systems. The heavy-duty equipment for these systems will be staged in the parking lot. The old equipment will come off the roof and remain in the parking lot until the disposal crew picks it up. The new equipment will be dropped off in the parking lot, and a crane will lift it to the roof when the workers are ready. You scheduled this project to begin on June 15 because all the students and faculty are gone for the summer. You list this assumption in the project charter. During the kickoff meeting, one of your stakeholders informs you that the school building is occupied during the summer. A neighboring community college uses the building (and the parking lot) to hold classes. You will add the community college as a stakeholder on the project and devise an alternative solution for equipment staging as you develop other planning documents.

High-Level Constraints

Constraints are anything that either restricts or dictates the actions of the project team. For example, you may have a hard due date that can't be moved. If you're developing a trade show event that occurs on September 25, this date is a constraint on the project because you can't move it. Budgets, technology, scope, quality, and direct orders from upper management are all examples of constraints.

The term *triple constraint* is one you'll hear often in project management circles. The triple constraints are time, scope, and cost, all of which affect quality.

High-Level Risks

Risks pose either opportunities or threats to the project. Most of the time, we think of risks as having negative impacts and consequences.

You should include a list of high-level risks in the project charter. These may cover a wide range of possibilities, including budget risks, scheduling risks, project management process risks, political risks, legal risks, management risks, and so on. The difference between a risk and a constraint is that a constraint is a limitation that currently exists. A risk is a potential future event that could impact the project. Beginning the project with a hard due date of September 25 is a constraint. The potential for a vendor missing an important delivery on September 15 is a risk.

Other Contents

Other elements you should describe in your charter include the name and authority level of the project manager, the name of the project sponsor, and any team members you've committed ahead of time to serve on the project team.

Criteria for Approval

Last but not least, the project charter should outline the criteria for project approval. Think of this as the definition for a successful project. These criteria will be used to determine whether the deliverables and the final product, service, or result of the project are acceptable and satisfactory. They might include items such as quality criteria, performance criteria, fitness for use, and so on. This section should also describe the process that stakeholders will use to indicate their acceptance of the deliverables.

Formal Approval

The project sponsor should review and sign the project charter. This sign-off provides the project manager with the authority to move forward, and it serves as the official notification of the start of the project. This approval is usually required prior to the release of purchase orders or the commitment by functional managers to provide resources to support the project.

Issuing the project charter moves the project from the Initiating phase into the Planning phase. Make certain all of the stakeholders receive a copy of the charter or can access it from your project repository. It is also a good idea to schedule a meeting to review the charter, review the next steps, and address any questions or concerns they may have.

Holding the Kickoff Meeting

Once the project charter is signed and approved, your next task is to hold a project kick-off meeting. This meeting should include the sponsor, your key project team members, and the key stakeholders on the project. You'll want to address and discuss most of the sections in the charter during this meeting. It's important that everyone understands the goals and objectives of the project, the project description, the high-level milestones, and the general project approach. This document and the project scope statement, covered in Chapter 4, are the two documents you'll come back to when stakeholders try to steer you or the team in a different direction than what was originally outlined. I'm not saying that stakeholders would ever do this on purpose, but trying to sneak in one more feature or making this "one little change" tends to make its way into most projects I've worked on. These documents are your safety net and the way to keep out-of-control requests at bay. The project sponsor signs the charter, and the key stakeholders on the project sign the project scope statement. So, they can't say they didn't know!

Chapter 6, "Resource Planning and Management," will talk more in-depth about the project kickoff meeting.

 Real World Scenario

Main Street Office Move

After your meeting with Kate, you start writing the project charter. Here is a rough draft outline of that document for the Main Street Office Move:

Purpose for the Project   To bring together all employees into one building. This will improve communication, reduce travel costs, improve productivity, and reduce lease costs.

Project Goals   This project will relocate 1,200 employees from three locations to the Main Street Office Building location. The move will be completed by December 31. All employees will report to work on January 2 at the Main Street Office Building.

Project Description   Relocate 1,200 employees to one location at the Main Street Office Building. The move is planned over the holiday period so that it's as minimally disruptive as possible. All employees have the period from December 25 to January 1 as paid time off. All employees will pack their personal belongings and take with them when they leave for the holiday period. All work-related items will be packed before leaving the evening of December 24.

Key Deliverables   Communication to all employees regarding the purpose for the move and instructions regarding packing belongings, location of new office, and timelines. Communication should occur in multiple forms at least five times before the move.

Procure the services of a moving company to move boxes and furniture. Procurement process to start six months prior to the move date.

Survey the Main Street Office Building and work with functional managers to determine seating charts.

Procure new office furniture. Delivery and setup take place between December 26 and December 31.

High-Level Requirements   Communication methods include all-hands meetings at each existing office location to explain the move. Establish a wiki site on the intranet with details as they become available.

The moving company chosen to perform this move will have experience in relocating offices.

All furniture will conform to the chosen color and pattern theme.

High-Level Milestones

- Communication to employees completed

- Moving company procurement completed

- Seating charts approved and finalized

- New furniture delivered and placed by December 31

- Office 1 move completed

- Office 2 move completed

- Office 3 move completed

High-Level Budget   The total budget for this project is $450,000.

High-Level Assumptions

- Employees are supportive of the move.

- Moving companies are available during the move week.

- Employees will have personal belongings packed and removed from premises prior to move.

- Furniture is delivered on time.

High-Level Constraints

- December 31 completion date

- $450,000 budget

High-level Risks   Bad weather during the move week

Name of the Project Manager   You. Your authority level consists of managing the budget, expending funds to perform the work, requesting resources to assist you, and contracting with a moving company for services.

Criteria for Project Approval   The project will be considered successful when all three offices are relocated by December 31 and it stays within the $450,000 budget.

Summary

Collectively, project management consists of five process groups: Initiating, Planning, Executing, Monitoring and Controlling, and Closing. Each of these process groups consists of individual processes that each have inputs, tools and techniques, and outputs.

Initiation is the formal authorization for the project to begin. It starts with a project request that includes the business case (which in turn includes the purpose or justification for the project) and outlines the high-level scope definition. The output from the Initiating process is the project charter. This document becomes the basis for more-detailed project planning. It should contain the purpose or justification for the project, project goals, project description, high-level requirements, high-level milestones, high-level budget, assumptions, constraints, high-level risks, name of the sponsor, name of the project manager, and criteria for approval.

Planning is the process where many project processes and documents are created, including the work breakdown structure, project schedule, budget, change management plan, communication plan, and more.

Monitoring and Controlling involves monitoring the performance of the project to ensure that deliverables meet quality deliverables, to make certain risks are in check, to assure changes follow the proper processes, and to control the project costs.

Closing is where the project is accepted and formal sign-off occurs. The final product, service, or result of the project is handed off to other areas of the organization to maintain going forward. Lessons learned are documented here, resources are released to their functional areas, and contracts are closed out.

The project charter provides formal approval for the project to begin and authorizes the project manager to apply resources to the project. The project sponsor is the one who publishes, signs, and approves the project charter.

Exam Essentials

Be able to define the Initiating phase.   Initiation authorizes the project to begin.

Be able to define the Planning phase.   This process is where most project documents and processes are created, including the schedule, work breakdown structure, budget, communication plan, procurement plan, and more. These documents are used as the foundation for managing the project throughout the remaining processes.

Be able to define the Executing phase.   The work of the project is performed in the Executing process. This is where the deliverables are produced.

Be able to define the Monitoring and Controlling phase.   This process monitors and controls the work, deliverables, and outputs of the project to determine whether there are variances from the project plan. Corrective actions are taken during this process to get the project back on course. Risks, issues, quality assurance, changes, and budget are among the elements of the project monitored during this process.

Be able to define the Closing phase.   Closing is where the product, service, or result of the project is accepted and formal sign-off occurs. Lessons learned are documented, resources are released, and contracts are closed out.

Be able to describe a project charter and list the key components.   A project charter provides formal approval for the project to begin and authorizes the project manager to apply resources to the project. The key components are the purpose or justification for the project, project goals and objectives, project description, key deliverables, high-level list of requirements, high-level milestones, high-level budget, high-level assumptions, high-level constraints, high-level list of risks, name of the sponsor, name of the project manager, and criteria for project approval.

Key Terms

Before you take the exam, be certain you are familiar with the following terms:

assumptions

Closing

constraints

deliverables

Executing

Initiating

iterative

milestones

Monitoring and Controlling

Planning

project charter

project description

requirements

triple constraints

Review Questions

1. The Initiating phase includes which task?
    A.   Assigning work to project team members
    B.   Sequencing project activities
    C.   Approving a project and authorizing work to begin
    D.   Coordinating resources to complete the project work

2. This person is responsible for authorizing the project to begin and signing the project charter.
    A.   Project sponsor
    B.   Executive in the organization who requested the project
    C.   Project champion
    D.   Project manager

3. Quality assurance, performance measuring and reporting, and change control are all part of which process?
    A.   Closing
    B.   Planning
    C.   Executing
    D.   Monitoring and Controlling

4. A primary role of the project manager includes informing this person of changes, status, conflicts, and issues on the project.
    A.   The project requestor
    B.   The project sponsor
    C.   The project champion
    D.   The most influential project stakeholder

5. Which of these activities occur during the Planning phase? Choose three.
    A.   High-level scope definition
    B.   Budget
    C.   Business case
    D.   Schedule
    E.   Change control
    F.   Resources

6. The Closing phase addresses which of the following? Choose three.
    A.   Project sign-off
    B.   Lessons learned
    C.   Governance processes

    D. Change control closeout

    E. Integration plan is put into place.

    F. Deliverables verification

7. Which of the following options are processes in the project management process groups? Choose three.

    A. Risk

    B. Initiating

    C. Monitoring and Controlling

    D. Procurement

    E. Scope

    F. Planning

8. This phase is where the work of the project is performed.

    A. Planning

    B. Monitoring and Controlling

    C. Initiating

    D. Executing

9. Which of the following is true concerning the project charter?

    A. Describes the project schedule

    B. Contains cost estimates for each task

    C. Authorizes the start of the project work

    D. Lists the responsibilities of the project selection committee

10. Which of the following is performed once the project charter is signed?

    A. You should hold a project kickoff meeting.

    B. You should write the project scope statement.

    C. You should submit the request and the project charter to the project selection committee.

    D. You should develop the project schedule.

11. You have just defined the major events for the project that will be used to determine and measure checkpoints throughout the project and determine whether the project is on time. What are they?

    A. Deliverables

    B. Goals

    C. Milestones

    D. Tasks

12. Your project sponsor has just signed the project charter. You held the kickoff meeting, and everyone on the team is anxious to get started on the work of the project. You are having a tough time holding them back and tell them you need to develop the work breakdown structure, schedule, and communication plan as a start. Which of the following is true regarding this scenario?

   A. You've just completed the Initiating process.

   B. Your team wants to jump right to the Executing process.

   C. The next step in the project lifecycle is to begin the Planning process group.

   D. The project sponsor has approved resources and funding for the project.

   E. All of the above.

13. Identify the items that should *not* be included in a project charter. Choose three.

   A. High-level budget

   B. Project objectives

   C. High-level cost-benefit analysis

   D. Equipment and resources needed

   E. Business case

   F. Project description

   G. High-level list of risks

14. Which of the following describes the difference between deliverables and milestones? Choose two.

   A. Milestones are used to measure performance.

   B. Deliverables are used to measure performance.

   C. Milestones are an output or result that must be completed in order to consider the project complete.

   D. Deliverables are an output or result that must be completed in order to consider the project complete.

15. These items are developed in the Initiating phase. Choose three.

   A. Project budget

   B. Procurement plan

   C. Scope statement

   D. Business case

   E. High-level scope definition

   F. High-level risks

16. Randy is a key technical resource for your project. You've worked with Randy on past projects and have identified him as one of the team members who will work on the project. The charter has been published, and there is great excitement about this project. You've scheduled a meeting to talk to Randy's functional manager next week. Which of the following conditions does this describe?

A. Risk

B. Assumption

C. Deliverable

D. Constraint

17. From the following options, select those that best describe the definition of a deliverable. Choose three.

A. Marks the completion of a project phase

B. Has measurable outcomes or results

C. Is a specific item that must be produced to consider the project complete

D. Describes detailed characteristics

E. Is documented in the business case

F. Is tangible and easily verified

18. Which of the following does *not* describe a constraint?

A. Project team actions are dictated.

B. It may regard budget, resources, or schedule.

C. Project team actions are restricted.

D. Project situations are believed to be true.

19. This component of the project charter describes the characteristics of the product of the project.

A. Milestones

B. Deliverables

C. Project description

D. Goals and objectives

20. Your project is to be performed outdoors. You are only four days from the big event, and there is a hurricane headed for shore. This is an example of which of the following?

A. Risk

B. Assumption

C. Deliverable

D. Constraint

Creating the Work Breakdown Structure

THE COMPTIA PROJECT+ EXAM TOPICS COVERED IN THIS CHAPTER INCLUDE

✓ **1.6 Given a scenario, execute and develop project schedules.**

 - Work breakdown structure

✓ **2.1 Given a scenario, predict the impact of various constraint variables and influencers throughout the project.**

 - Common constraints
   - Budget
   - Scope
   - Deliverables
   - Quality
   - Environment
   - Resources
   - Requirements
   - Scheduling
 - Influences
   - Change request
   - Scope creep
   - Constraint reprioritization
   - Interaction between constraints
   - Stakeholder/sponsors/management
   - Other projects

✓ **4.1 Compare and contrast various project management tools.**

 - Charts
   - Process diagram

Now that you have an approved project charter, it is time to talk about project planning. In many cases, I've seen that once a project is approved, people want to start working on the project activities immediately. The general consensus is, "Who has time for planning? We need to get this project started!"

As the project manager, it's your responsibility to write the project plan and make certain everyone, including team members and key stakeholders, understands that plan. The project coordinator will assist you in preparing the plan, in cross-functional coordination with the business areas that will participate on the project, and with preparing time and resource schedules.

The first planning topic discussed is the project scope document.

Project scope includes all the components that make up the product or service of the project and the results the project intends to produce. Scope planning will assist you in understanding what's included in the project boundaries and what is excluded.

You'll need to define and document three scope components to complete scope planning: the scope management plan, the scope statement, and the work breakdown structure (WBS). The scope management plan documents how the project scope will be defined and validated and how scope will be monitored and controlled throughout the life of the project. The scope statement provides a common understanding of the project by documenting the project objectives and deliverables. The final component of scope is the work breakdown structure, which breaks down the project deliverables into smaller components from which you can estimate task durations, assign resources, and estimate costs.

Documenting the Scope Management Plan

The *scope management plan* describes how the project team will define project scope, validate the work of the project, and manage and control scope. The scope management plan should contain the following elements:

- The process you'll use to prepare the scope statement

- A process for creating, maintaining, and approving the work breakdown structure

- A definition of how the deliverables will be validated for accuracy and the process used for accepting deliverables

- A description of the process for controlling scope change requests, including the procedure for requesting changes and how to obtain a change request form

One of the most important elements of the scope management plan, the change request process, will help master that dreaded demon that plagues every project manager at one time or another—*scope creep*. Scope creep involves changing the project or product scope without having approval to do so and without considering the impacts that will have on the project schedule, budget, and resources. It is the term commonly used to describe the changes and additions that seem to make their way into the project to the point where you're not managing the same project anymore. Scope creep usually occurs in small increments over time. A small change here, a small new addition there, and the next thing you know the project's overall objectives have changed. The solution, of course, for preventing scope creep is having a change request plan.

Scope change is inevitable on most projects. The key to dealing with scope change is describing how you'll handle it within the scope management plan.

If the project team defines the basic scope management framework early in the Planning processes, each team member has a point of reference to communicate with stakeholders who may come to them with "something they forgot to mention" when the scope statement was approved. Everyone involved in the project needs to understand that the rules set up during implementation need to be followed to make any request to change the scope of the project. Without a documented plan, you will soon find that interested parties are talking to team members directly and changes are happening outside of your control. The team members will, understandably, want to try to accommodate the customer's needs. But without analysis of the impact of these changes, adding 10 or 20 minor scope changes may put your schedule or budget or both in jeopardy.

 You may not encounter a lot of questions on the exam regarding the scope management plan. However, it is an important element in building the scope statement, controlling scope creep, and preparing the work breakdown structure. It's also a component of the scope baseline that consists of the scope management plan, the scope statement, and the work breakdown structure.

Writing the Scope Statement

The *scope statement* includes all the components that make up the product or service of the project and the results the project intends to produce. Although this sounds simple and straightforward, a poorly defined scope statement can lead to missed deadlines, cost overruns, poor morale, and unhappy customers. Good scope planning helps ensure that all the work required to complete the project is defined, agreed on, and clearly documented.

The scope statement builds on and adds detail to the high-level scope definition you documented in the project charter. Depending on the detail of work completed during the project Initiating process, the scope statement may also include a more detailed analysis of the product, an additional cost-benefit analysis, and an examination of alternative solutions.

You may also find that much of the work required for scope planning was already completed and documented in the business case and project charter documents. If that's the case, congratulations—you are now ahead of the game.

 The processes to define the scope elements are iterative; that is, you will continue to define and refine the project scope (and other planning elements), going back over them several times until you and the team are satisfied everything has been identified and documented.

The purpose of the scope statement is to document the project objectives, the deliverables, and the work required to produce the deliverables. It is then used to direct the project team's work during the Executing processes and as a basis for future project decisions. The scope statement is an agreement between the project team and the project customer that states precisely what the work of the project will produce. Simply put, the scope statement tells everyone concerned with the project exactly what they're going to get when the work is finished. Any major deliverable, feature, or function that is not documented in the scope statement is not part of the project. There isn't a hard and fast rule on what to include in the scope statement. It can be as detailed as needed depending on the complexity of the project. Typically, the scope statement includes the project objectives, a project description, acceptance criteria, key deliverables, success criteria, exclusions from scope, time and cost estimates, project assumptions, and constraints. Chapter 3 covered some of these items in detail. Let's look at the others next.

 You'll find a sample scope statement, including most elements discussed next, in the chapter's final case study.

Project Objectives

Objectives describe the overall goal the project hopes to achieve. Objectives should be measurable and verifiable, and they are often time-bound. For example, you may have a final completion date for the entire project or for some of the project's key objectives. You can reuse the goals and objectives you documented in the project charter. If you've learned new or more detailed information about the objectives since writing the charter, be sure to include that here.

Project Description

The project scope description describes the key characteristics of the product or service you are creating through this project. Again, you could reuse the project description you documented in the project charter and add more details to it in this document.

If the result of your project is a tangible product, you should include the product scope description here as well. The *product scope description* describes the features, functions, and explains the major characteristics of the product.

Acceptance Criteria and Key Performance Indicators

Acceptance criteria include the process and criteria you'll use to determine that the deliverables are complete and satisfactorily meet expectations.

The final acceptance criteria describe how you'll determine whether the entire project is complete and meets expectations.

Key performance indicators (KPIs) help you determine whether the project is on track and progressing as planned and whether deliverables meet expectations. KPIs are monitored periodically and alert you that you must take action to get the project back on track. An example of a KPI might be specifying that a deliverable must meet certain performance throughput measures.

Key Deliverables

Deliverables, as you recall, are measurable outcomes, measurable results, or specific items that must be produced to consider the project or project phase completed. Deliverables identified in the scope document should be specific and verifiable.

Critical Success Factors

Deliverables and requirements are sometimes referred to as *critical success factors*. Critical success factors are those elements that must be completed accurately and on schedule in order for the project to be considered complete. They are often key deliverables of the project, and if their description in the scope document is not accurate or complete, they will likely cause project failure.

Exclusions from Scope

Exclusions from scope are anything that isn't included as a deliverable or work of the project. It's important to document exclusions from scope so there is no misunderstanding about features or deliverables once the product is complete.

Time and Cost Estimates

Depending on the organization, you may come across scope statement templates that require time and cost estimates. In this section, you'll provide an estimate of the time it will take to complete all the work and the high-level estimates for the cost of the project. These will be *order-of-magnitude* estimates based on actual duration and cost of similar projects or the expert judgment of someone familiar with the work of the project. Order-of-magnitude estimates are usually wide-ranging and do not have to be precise estimates at this stage in the project.

Assumptions

You'll recall from Chapter 3 that an assumption is an action, a condition, or an event that is believed to be true. Assumptions can get you into trouble if they are not documented and clearly understood by the stakeholders and project team members. You may think something is obvious, but if it's not written down, chances are other team members or stakeholders will have a different opinion on the matter. Assumptions must be documented and validated.

Constraints and influencers are the last element of the scope statement. They are topics unto themselves, so you'll take a look at them in the next section.

Scope Statement Purpose

Remember for the exam that the purpose of the scope statement is to document the project objectives, the deliverables, and the work required to produce the deliverables.

Constraints and Influences

The last section of the scope statement is the constraints and influences list. Remember that a constraint is anything that restricts or dictates the actions of the project team. An influence can bring about a constraint or impact an existing constraint. Every project faces potential constraints regarding time, budget, scope, or quality. From the start of any project, at least one of these areas is limited. If you are developing a new product with a short time-to-market window, time will be your primary constraint. If you have a fixed budget, money will be the constraint. If both time and money are constrained, quality may suffer.

A predefined budget or a mandated finish date needs to be factored into any discussion on project scope. Scope will be impacted if either time or budget is constrained. As the project progresses and changes to scope are requested, scope may become a constraint that in turn drives changes to time, cost, or quality.

Project constraints and influences comprise 17 percent of the exam questions. Make certain you understand the definition of a constraint and influence and also understand the types of constraints and influences that exist on any project.

Constraints

According to CompTIA, the following are common constraints found on many projects:

- Budget
- Scope
- Deliverables

- Quality
- Environment
- Resources
- Requirements
- Scheduling

   You'll look at each of these next.

Budget

I have never had the experience of working on a project where budget was *not* a constraint. All projects have a limited amount of funding available to perform the work or to purchase the services required to complete the project. The project costs must be monitored and controlled throughout the project so that you stay within budget. Work with your sponsor and key stakeholders to determine the project budget as early as possible in the Planning process.

Scope

Earlier in the chapter you discovered that scope describes the project deliverables and outlines the expectations and acceptance criteria for the deliverables and for a successful project. Scope is a constraint because it dictates the actions of the project team in relation to fulfilling the deliverables. If you do work other than what's required to fulfill the deliverables outlined in the scope statement (yes, the dreaded scope creep problem), you have a runaway project on your hands and won't likely have a successful outcome.

   If scope is not fully understood, defined, and documented, you will find that time or budget or both are also impacted.

Deliverables

Deliverables, as discussed earlier, are measurable outcomes, measurable results, or specific items that must be produced to consider the project (or project phase) completed. Deliverables should be specific and verifiable. Deliverables are constraints because the specific requirements or measurable results drive (or restrict) the actions of the project team.

Quality

Quality concerns measuring or quantifying performance, deliverables, functionality, specifications, and so on. Quality assurance is defined during the Planning process and measured and controlled throughout the project. For example, if a quality standard requires deliverables to weigh 9 ounces or more, a deliverable weighing 7.5 ounces will not meet quality standards.

Environment

The environment can be a constraint in any number of circumstances. Weather is an environmental factor that can be a constraint, as are the rugged conditions of the Australian Outback or Antarctica. Environmental constraints could also include air-quality or water-quality standards, for example, or emissions regulations. It's important to understand any environmental factors that may restrict or dictate the actions of the team.

Resources

Resources can range from human resources to materials to equipment to funding and more. Resources can be a constraint when they are scarce, have limited availability, or cannot be delivered on time. Your organization may not have the funding, technology, equipment, or human resources with the skill sets needed to fulfill the deliverables in the time frame required; therefore, resources become a constraint.

Requirements

Requirements describe the characteristics of the goals or deliverables that must be met in order to satisfy the needs of the project. Requirements might also describe results or outcomes that must be produced in order to satisfy the deliverables as documented in the scope statement.

Scheduling

Schedule is another constraint that exists on virtually all projects. As with budgets, I've never had the privilege of working on a project that didn't have a time constraint. The constraint can take a couple of forms. It could be a due date set by your executive management. It could also be driven by forces external to the project. For example, the summer Olympic Games must be held during the summer months. Perhaps a resource you need for your project is unavailable during the months of April, September, and October. Scheduling then becomes a constraint because you must work within their availability window to complete the deliverable for your project. Again, work with your project sponsor and key stakeholders to determine as early as possible in the project what their expectations are regarding deliverable due dates, project completion dates, and other time frames or dates that may be off-limits.

You do not have the luxury of working on a project with no constraints. Remember that the most common constraints on any project are scope, budget, and time. You will need to balance and weigh constraints against each other to determine the best way to accomplish the goals of the project while also deciding which constraint is more important and which constraint has more flexibility. This requires communication with your sponsor and stakeholders and managing expectations.

Influences

Influences are those factors that may impact or change an existing constraint or may bring about a new constraint. According to CompTIA, the following are common influences:

- Change request
- Scope creep
- Constraint reprioritization
- Interaction between constraints
- Stakeholder/sponsors/management
- Other projects

You'll look at each of these next.

Change Request

A change request can impact existing constraints or bring about a new constraint on the project. For example, if your project is already time-constrained and the change request your stakeholders just approved changes the project due date, the scheduling constraint is impacted.

Scope Creep

I've talked a lot about scope creep. It is imperative that you control changes to scope with an established change management procedure to assure project success. You'll learn more about change management processes in Chapter 9.

Constraint Reprioritization

Reprioritizing the constraints may change their impacts or influence on the project. For example, let's say the primary constraint on your project is scheduling because you have a due date that is required by the project sponsor for the project. Let's also say the company is undergoing some financial difficulties and you've just learned that the budget for your project has been reduced. The budget reduction is so significant it now becomes the primary constraint on your project.

Interaction Between Constraints

An example of interaction between constraints may occur when you have changes to scope that in turn impact the schedule and/or budget. Scope can change through the formal change control process or through scope creep. Perhaps your project sponsor has decided you need one more deliverable in order to satisfy the overall goals and objectives of the project. This is approved in a formal change request process. However, scope changes require schedule changes, and they often require budget changes as well. The interaction here is like a domino effect, and all the constraints should be reexamined to ensure that they still accurately reflect the conditions of the project. This interaction may also cause a reprioritization of the constraints.

Stakeholders, Sponsors, and Management, Oh My

Project sponsors and executive management are notorious for changing project priorities on a moment's notice. Today they want deliverable A; tomorrow they want Deliverable Z. It's a constantly changing target! They also sometimes lose interest in the project. Newer projects may gain in importance in the organization (or wane), causing changes to your project that could bring about new constraints or change existing ones. For example, if your project is time-constrained and another project comes along that rises in importance, you may lose resources or budget to the new project, which will bring about new constraints on your project. In my experience, the single biggest influence on project constraints is sponsors and executive management changing their priorities.

 Real World Scenario

A Sample Scope Statement

You have been asked to set up a fundraising project for your school. Here is a sample of what your scope statement might look like:

Project Objective (from the Project Charter)   Establish a fundraising golf tournament to raise $20,000.

Project Description   Hold a golf tournament the last week of the school year to raise funds for classroom equipment and resources. Our target goal is $20,000 in donations.

Acceptance Criteria   Golf tournament raises $20,000. Participants enjoy the tournament and recommend it to their friends next year.

Key Deliverables   The major deliverables are as follows:

- Establish two major sponsors (potentially corporate sponsors) willing to contribute $5,000 each.

- Establish multiple minor sponsorships between $500 and $1,000 that will total another $5,000 at a minimum.

- Establish pricing for golf tournament participants that will cover green fees, cart fees, and provide $50 toward the fundraiser.

- Reserve a golf course for the tournament.

- Devise challenges for participants such as longest drive, longest putt, and best scores.

- Procure prizes.

- Develop marketing materials to advertise the event.

- Train volunteers to work at the event.

Exclusions from Scope   Alcohol is available at the golf course but is not included in the entrance fees and must be purchased separately.

Time and Cost Estimates   The time estimate is one FTE for a full year to coordinate the event and procure sponsors, prizes, and the golf course agreement. Costs are minimal as sponsors will be providing the funds to purchase prizes.

Assumptions

- Golf courses are available the day of the tournament.

- The number of sponsors and funding needed will be obtained. Players will attend the tournament, and all available spaces will be filled.

Constraints   The date of the tournament is within one week of the last week of school.

Approval of the Scope Statement

Once you have completed the scope statement, your next step is to conduct a review session with your project team to make sure that everyone is in agreement and there are no unresolved issues or missing information.

The next step is to present the scope statement to all the stakeholders, including the project sponsor and the customer. Attach a sign-off and approval sheet to the back of the scope statement with enough signature lines for the sponsor and each of the major stakeholders on the project. Their approval on this document assures their buy-in regarding the scope of the project and should be required before any project work is undertaken.

If you've defined the scope of the project, gained stakeholder approval, and have gotten all the major stakeholders and the sponsor to sign the scope statement, you're well on your way to a successful project outcome. Taking the time to create a well-documented scope statement will also help in establishing a solid basis for future change management decisions.

Documenting the Requirements

You may recall from Chapter 3 that requirements describe the characteristics of the goals or deliverables that must be met in order to satisfy the needs of the project. Requirements might also describe results or outcomes that must be produced in order to satisfy a contract, specification, standard, or other project document. Requirements quantify and prioritize the wants, needs, and expectations of the project sponsor and stakeholders.

Requirements definition can be part of the scope statement, or it can be an independent document, depending on the size and complexity of the project. In my experience, the scope statement works fine to document the requirements for small projects.

You will take a closer look at requirements next.

Requirement Categories

Requirements fall into several categories. Having a good understanding of the differences in requirements can help you when writing a requirements document. Your stakeholders won't know the difference and will mix business requirements with functional and non-functional requirements when discussing their expectations. On larger projects it helps to categorize the requirements so when you're constructing the work breakdown structure and later assigning resources, they will already be somewhat organized.

Business Requirements

An organization's *business requirements* are the big-picture results of fulfilling a project and how they satisfy business goals, strategy, and perspective. Business results can be anything from a planned increase in revenue to a decrease in overall spending to increased market awareness and more.

When gathering requirements, your focus should be on the "what," not on the "how." Stakeholders are passionate about their needs and will likely have a list of ideas on "how" to solve the problem and implement the project. You want to drive them to the "what" and to answer the question, "What problem are we trying to solve?"

When documenting business requirements, sometimes it's helpful to use a process diagram. This shows step-by-step how a process works, where approvals or decisions need to be made, and so on. Process diagrams also come in handy when mapping out business processes. For example, the PMO may have a process diagram that documents how a project idea turns into a project. It starts with a project idea, goes to the selection committee, and returns to the PMO to either proceed into a full-fledged project or for the idea to be archived or placed in a hold status. Figure 4.1 shows a simple process diagram.

Functional Requirements

Functional requirements are the product characteristics needed for the product to perform. They are typically behavioral in nature or performance oriented and may also describe elements such as color, quantity, and other specifications. For example, if you're designing a new toaster, a functional requirement might look like this: pushing the green button will pop the toast out of the toaster, and pushing the red button will begin the toasting process.

Industry or corporate standards may impact your functional requirements. In our toaster example, perhaps industry standards state that electric cords on kitchen appliances must be 20 inches or shorter in length. This requirement needs to be added to the requirements document. You'll need to research any industry standards that may impact your requirements and document them because they may impact activity duration or cost estimates.

FIGURE 4.1   Sample process diagram

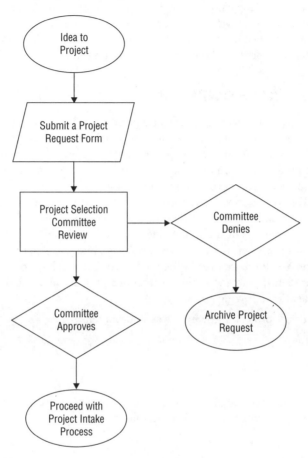

NOTE

If you work in a regulated industry, make sure you address the question of whether any specific government or industry-related regulations impact the design or delivery of your product. Regulatory noncompliance is a serious offense, and correcting infractions after the fact can be both time-consuming and costly.

Non-functional Requirements

Non-functional requirements describe the characteristics of the functional requirements. They are not performance or behavioral based. For example, a non-functional requirement in a toaster example might state that the green button should be 1.5 inches in diameter and the red button should be 2 inches in diameter.

In my experience, stakeholders are typically more prepared to discuss functional requirements than non-functional, so come prepared with a list of questions. If all else fails, asking why and what always works. "Why does the green button stop the toaster and pop the toast?" And, "What would happen if I pushed the green button and the red button at the same time?" If your PMO has a requirements template or checklist, use it in your meetings with the client group.

The Requirements Document

As I mentioned earlier, requirements quantify and prioritize the wants, needs, and expectations of the project sponsor and stakeholders to achieve the project objectives. Requirements typically start out high level and are further defined, or progressively elaborated, as the project progresses. You must be able to track, measure, test, and trace the requirements of the project. You never want to find yourself at the end of the project and discover that you have no way to validate the requirements. If you can't measure or test whether the requirement satisfies the business need of the project, the definition of success is left to the subjective opinions of the stakeholders and team members.

You've worked hard to gather and define requirements, and you don't want all that effort going to waste. You'll record everything you've learned in a requirements document or in the scope statement. If your requirements document is a stand-alone document, you should include at least the following elements:

- Business need for the project and why it was undertaken
- Project objectives
- Project deliverables
- Requirements

Once you have documented the project scope and requirements, you're ready to start on the work breakdown structure. You'll look at that next.

Creating the Work Breakdown Structure

The final element of scope planning is the *work breakdown structure (WBS)*. The WBS is a deliverables-oriented hierarchy that defines all the work of the project. Each level of the WBS is a further breakdown of the level above it. *Decomposition* is the process of breaking down the high-level deliverables (and each successive level of the WBS) into smaller, more manageable work units. Once the work is broken down to the lowest level, you can establish time estimates, resource assignments, and cost estimates.

A WBS is one of the fundamental building blocks of project planning. It will be used as an input to numerous other planning processes. It's also the basis for estimating activity duration, assigning resources to activities, estimating work effort, and creating a budget. Because the WBS is typically displayed as a graphical representation, it can be a great way of visually communicating the project scope. It contains more details of the deliverables than the scope statement does and helps further clarify the magnitude of the project deliverables.

The WBS puts boundaries around the project work because any work not included in the WBS is considered outside the scope of the project.

Decomposing the Major Deliverables

The quality of your WBS depends on having the right team members involved in its development. You won't want a large team of people to assist in this process, but it's helpful if you can involve some of your more experienced team members. Work with the functional managers to get representation from each business unit that has a major deliverable for the project.

A WBS is typically created using either a tree structure diagram or an outline form. The tree structure can be created using software, using a whiteboard, or using easel paper with sticky notes for each level and each component of the WBS. This allows the components to be moved around as you work through the process and get everything in proper order.

A typical WBS starts with the project itself at the topmost level. The next level consists of the major deliverables, project phases, or subprojects that support the main project. From there, each deliverable is decomposed into smaller and smaller units of work. The lowest level of any WBS is called the *work package level*. This is the level where resources, time, and cost estimates are determined. Work packages are assigned to team members or organizational units to complete the activities associated with this work.

Make sure that all the participants have reviewed the project charter, scope statement, and requirements document and have a clear understanding of all the deliverables. Have copies of these documents available as a reference. The team may go through several iterations of constructing the WBS before it's considered complete.

Figure 4.2 is an abbreviated example of a WBS for a conference event project.

FIGURE 4.2   Sample WBS

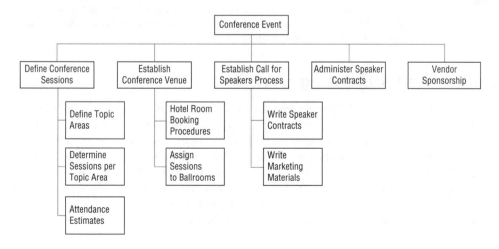

The second method for depicting a WBS is the outline form. Regardless of how you depict the WBS, each level in the WBS should have a unique identifier. This identifier is typically a number, and it's used to sum and track the costs, schedule, and resources associated with the WBS elements. These numbers are usually associated with the corporation's chart of accounts, which is used to track costs by category. Collectively, these numeric identifiers are known as the *code of accounts*. If you took the example from Figure 4.2 and converted it to outline form using the code of accounts, it would look like this:

Conference Event Project

1. Define conference sessions.

    1.1 Define topic areas.

    1.2 Determine number of sessions per topic.

    1.3 Determine approximate attendance at each session.

2. Establish conference venue.

    2.1 Research available hotel and conference space and reserve conference dates.

    2.2 Define hotel room booking procedure.

    2.3 Assign sessions to ballrooms.

3. Establish a call-for-speakers process.

    3.1 Write speaker contracts.

    3.2 Write marketing materials.

4. Administer speaker contracts.

5. Contact and sign up vendors for sponsorships.

> The project description, project scope, deliverables, and work breakdown structure will be *progressively elaborated* throughout the planning phase. As more information becomes available and as you decompose deliverables, you may discover elements of scope or the deliverables that need additional detail or clarification.

Guidelines for Creating a WBS

Getting started creating a WBS can sometimes seem overwhelming. Once you start breaking down your deliverables into smaller units of work, it's tempting to immediately put these work units in logical order and start assigning team members to them. You'll need to keep reminding the group that the purpose of the WBS is to make certain each deliverable is identified (usually as a level-2 element) and that each subdeliverable is decomposed from there. This isn't the process where activities are put into logical order or where predecessor or successor work is determined. That will occur when the project schedule is developed. For now, focus on the deliverables and their decomposition.

Although there is no one right way to complete a WBS, there are some tips you can use to help be more successful. Here are some helpful guidelines to review with the project team before diving into a WBS session:

Recruit knowledgeable resources.   Do not try to complete the WBS yourself in the interest of saving time. If you are not an expert on the deliverables, you will miss key elements. You also want the team members to assist in this process so that they "buy in" to the work of the project and can see the level of effort needed to bring about a successful conclusion. Involving knowledgeable team members in the creation of a WBS is far more effective at communicating what the project is about than handing someone a completed WBS.

Each item in a lower level is a component of the level above.   Completing all the items at the lower levels of a WBS leads to the completion of the higher-level components. As a checkpoint, you should review the items at the lower level and ask the team whether completion of those items will result in completion of the components of the next higher level. If the answer to this question is no, then you have not identified all the lower-level tasks.

Define the work package level.   Make sure you work the WBS to a level where the team feels comfortable that resources can be assigned and held accountable for completing the work and estimates can be determined. This is the primary goal of constructing the WBS. Sequencing, assigning resources, and estimating are all separate activities that you will perform after the WBS is complete.

Do not create a to-do list.   You should not decompose work components into individual activities. Otherwise, you will spend your entire project-management experience managing individual checklists and activities for the work packages. The person who is assigned to the work package level is the one responsible for determining and managing all the activities that make up the work package.

Use the appropriate number of levels.   Each major deliverable may have a different level of decomposition. It is not uncommon for one portion of the WBS to have three levels and another to have five levels. You should be concerned about getting to a manageable work package, not about balancing the WBS. If you try to force an even number of levels across all deliverables, you will end up with some deliverables that are not broken down in adequate detail and others that end up listing every minor activity to complete a simple task.

Benefits of the WBS

The WBS is often listed as one of the most important components of a successful project. As you will see in later chapters, the WBS is an input to numerous project management processes.

The WBS is an excellent tool for team building and team communication. A graphic representation of the major project deliverables and the underlying subcomponents allows team members to see the big picture and understand how their part in the project fits in. The direct link between a given work package and a major project deliverable can also help clarify the impact on individual team members. Additionally, as new resources are added to the project, the WBS can help bring these new team members up to speed.

 If your organization typically undertakes projects that are similar in size, scope, and complexity, consider using your existing WBS as a template for future projects. It's a great starting point for the new project and can help kick off the brainstorming session with the project team.

A detailed WBS will not only prevent critical work from being overlooked, but it will also help control change. If the project team has a clear picture of the project objectives and the map to reach these objectives, they are less likely to go down a path unrelated to the project scope. I don't mean to imply that a WBS will prevent change; there are almost always changes during the project life cycle. But a WBS will clarify that a request is a change and not part of the original project scope. The WBS is also useful when discussing staffing requirements or budgets.

The WBS is an excellent tool for communicating with customers and stakeholders. People don't always comprehend the magnitude of a project until they see the diagram of the project deliverables and the subcomponents required to reach the objectives of the project. Seeing the work of the project displayed on a WBS will help you convey to the project team and the stakeholders the need for communicating at all levels of the project, coordinating work efforts, and adhering to the project scope.

WBS Dictionary

There's one more component of the WBS, called the *WBS dictionary*. This is where the WBS levels and work component descriptions are documented. These are some of the elements you should list in the WBS dictionary:

- Code of accounts identifier
- Description of the work of the component
- Organization responsible for completing the component
- Resources
- Cost estimates
- Criteria for acceptance

All the WBS components should be listed in the dictionary. This serves as a reference for you and the team regarding the WBS and should be easily accessible to all project team members.

Real World Scenario

Main Street Office Move

After the kickoff meeting, you're ready to write the scope statement and the work break-down structure. Your scope statement contains the following elements:

Project Objectives   To bring together all employees into one building. This will improve communication, reduce travel costs, improve productivity, and reduce lease costs. This project will relocate 1,200 employees from three locations to the Main Street Office Building location. The move will be completed by December 31. All employees will report to work on January 2 at the Main Street Office Building.

Acceptance Criteria   The move will be made on time and on budget, with minimal disruption to employees. Employees will return to work on January 2, and their computers and phones will be available for use, the printers will be online, all office spaces will be completed (including pictures hung on the walls), and office kitchens will be fully stocked with beverages and snacks. All fleet cars will be moved to the new garage located in the new building and will be available for checkout on January 2.

Major Deliverables   The deliverables include the following:

- Communication to all employees regarding the purpose for the move and instructions regarding packing belongings, location of new office, and timelines. Communication should occur in multiple forms at least five times before the move.

- Procure the services of a moving company to move boxes and furniture. Procurement process to start nine months prior to the move date.

- Procure general contracting services and perform building remodel on the three floors occupied by Kate's organization.

- Survey the Main Street Office Building and work with functional managers to determine seating charts.

- Procure new office furniture. Delivery and setup between December 15 and December 31.

- Procure and install cubicle furnishings.

- Work with interior designer to maximize space and provide aesthetically pleasing work space and common areas.

- Install employee desktops, peripherals, and telephones.

- Install three networked printers on each floor.

- Relocate fleet cars to the new garage.

Exclusions from Scope   This project does not include moving the satellite office located 90 miles south of the three main campuses.

Time and Cost Estimates   The high-level budget for this project is $450,000.

The high-level timeline for this project is 12 months. Kickoff occurred December 27. Move will occur 12 months from that date.

Assumptions   You have the following assumptions:

- Employees are supportive of the move.

- Moving companies are available during the move week.

- Employees will have personal belongings packed and removed from premises prior to move.

- Furniture is delivered on time.

- All sites will provide reasonable access for installers and movers.

Constraints   December 31 completion date

$450,000 budget

You discovered new deliverables and a stakeholder who was missed during the initiating phase. You also changed one of the deliverables (procure moving company) to begin nine months prior to the move rather than six. The information technology (IT) team is integral to the success of this project as all employee workstations and phones must be set up and functioning before returning on January 2. The IT manager has been added to the stakeholder matrix and invited to all future meetings. Additional deliverables and assumptions have been added to the scope statement.

Work Breakdown Structure (WBS)   This list represents your abbreviated, sample WBS.

1. Communicate move purpose with employees.

    1.1 Define communication methods and channels.

    1.2 Write communication content.

    1.3 Publish communication content to each channel.

2. Procure moving company services.

    2.1 Research moving companies who specialize in office moves.

    2.2 Research reviews and referrals of at least three moving companies.

    2.3 Interview moving companies.

    2.4 Reserve move dates.

    2.5 Develop and approve contract with moving company.

---

3. Determine seating charts.

    3.1 Meet with functional managers to determine needs.

    3.2 Obtain building blueprints.

    3.3 Perform walk-through of the building.

4. Procure furniture and fixtures.

5. Complete interior design sessions.

6. Install technology hardware and software.

7. Install three networked printers per floor.

8. Relocate fleet cars from three locations to the new building garage.

Summary

Scope planning uses the output of the initiating phase, the project charter, to create the scope statement and the scope management plan. The scope management plan documents the process you'll use to prepare the scope statement and WBS, a definition of how the deliverables will be validated, and a description of the process for controlling scope change requests.

The scope statement is the basis for many of the planning processes and future change decisions. It is also the basis for setting the boundaries of the project with the customer and stakeholders. A scope statement includes the product description, key deliverables, success and acceptance criteria, key performance indicators, exclusions, time and cost estimates, assumptions, and constraints.

Requirements describe the characteristics of the deliverables. They might also describe functionality that a deliverable must have or specific conditions a deliverable must meet to satisfy the objective of the project. They are typically conditions that must be met or criteria that the product or service of the project must possess to satisfy the objectives of the project. Requirements quantify and prioritize the wants, needs, and expectations of the project sponsor and stakeholders. They are documented in the scope statement or in a stand-alone requirements document. Requirements categories include business, functional, and non-functional.

The work breakdown structure is created by taking the major deliverables from the scope statement and decomposing them into smaller, more manageable components. The breakdown continues through multiple levels until the components can be estimated and resourced. The lowest level of decomposition is the work package level. The WBS includes all the work required to complete the project. Any deliverable or work not listed on the WBS is excluded from the project. The WBS is a critical component of project planning. A WBS is the basis for time estimates, cost estimates, and resource assignments.

The WBS dictionary should list every deliverable and each of their components contained in the WBS. It should include a description of the component, code of account identifiers, responsible party, estimates, criteria for acceptance, and any other information that helps clarify the deliverables and work components.

Exam Essentials

Describe the purpose of a scope management plan.   A scope management plan documents the procedures for preparing the scope statement and WBS, defines how the deliverables will be verified, and describes the process for controlling scope change requests.

Understand the purpose of the scope statement.   The scope statement is the basis of the agreement between the project and the customer concerning what comprises the work of the project. It defines the deliverables and success criteria that will meet those objectives.

Be able to list the components of a scope statement.   A scope statement includes a project description, acceptance criteria, key deliverables, exclusions from scope, assumptions, and constraints. It could also contain a high-level time and cost estimate to complete the project.

Be able to define requirements.   Requirements describe the characteristics of the deliverables, or functionality that a deliverable must have, or specific conditions a deliverable must meet to satisfy the objective of the project.

Know how to define and create a work breakdown structure.   The WBS is a deliverable-oriented hierarchy that describes the work required to complete the project. The WBS is a multilevel diagram that starts with the project, includes the major deliverables, and decomposes the major deliverables into smaller units of work to the point where time and cost estimates can be provided and resources assigned.

Understand the levels in a WBS.   The highest level of the WBS is the project name. The major deliverables, project phases, or subprojects make up the next level. The number of levels in a WBS will vary by project; however, the lowest level of the WBS is a work package.

Describe a WBS dictionary.   The WBS dictionary describes each of the deliverables and their components and includes a code of accounts identifier, estimates, resources, criteria for acceptance, and any other information that helps clarify the deliverables.

Key Terms

Before you take the exam, be certain you are familiar with the following terms:

> acceptance criteria
>
> business requirements

code of accounts

critical success factors

decomposition

functional requirements

key performance indicators (KPIs)

non-functional requirements

order of magnitude

product scope description

scope creep

scope management plan

scope statement

WBS dictionary

work breakdown structure (WBS)

work package level

Review Questions

1. Which of the following is not a key component of scope planning?

    A.   Work breakdown structure (WBS)

    B.   Scope statement

    C.   Project charter

    D.   Scope management plan

2. The scope statement provides which of the following?

    A.   A basis for a common understanding of the project and for making future decisions regarding the project

    B.   A detailed list of all resources required for project completion

    C.   A schedule of all the key project activities

    D.   A process for managing change control

3. Which of the following is a characteristic of a WBS?

    A.   A cost center structure for the project that describes the work of the project and the costs per work component to complete the deliverables

    B.   A deliverables-oriented chart of the work of the project with assignments showing the project teams responsible for the work components

    C.   A deliverables-oriented structure that describes the detailed tasks required to complete the deliverables

    D.   A deliverables-oriented structure that defines the work of the project

4. Which of the following are components of a scope statement? Choose three.

    A.   General project approach

    B.   Project description

    C.   Assumptions and constraints

    D.   Exclusions

    E.   Stakeholder list

    F.   High-level milestones

    G.   Change request process

5. A WBS is created using a technique called *decomposition*. What is decomposition?

    A.   Matching resources with deliverables

    B.   Breaking down the project deliverables into smaller, more manageable components

    C.   Estimating the cost of each individual deliverable

    D.   Creating a detailed to-do list for each work package

6. What is the lowest level of the WBS?

    A. Work package

    B. Level 5

    C. Milestone

    D. Activities

7. Which of the following describes influences?

    A. Influences can impact, change, or create a new constraint.

    B. Scope creep is an example of an influence.

    C. Change request is an influence.

    D. Interaction between constraints is an example of an influence.

    E. All of the above

    F. A, B, D

8. Which of the following is not a benefit of a WBS?

    A. A WBS is an excellent tool for team building.

    B. A WBS helps prevent critical work from being overlooked.

    C. A WBS can become a template for future projects.

    D. A WBS can be used to describe how the deliverables will be validated.

9. All of the following are true regarding code-of-accounts identifiers except for which one?

    A. These are unique numbers for each component on the WBS.

    B. They are documented in the WBS dictionary.

    C. They are tied to the organization's chart of accounts.

    D. They are assigned to the resources who are associated with the work package level.

10. Your team has already created a WBS for the ABC product launch project. You are kicking off phase 2 of this project, which is the product development phase. Which of the following is an example of what might appear in the second level of your project's WBS?

    A. ABC Product Launch Project

    B. Project deliverables

    C. Project phases

    D. Activities

11. Which element is not a component or function of the scope management plan?

    A. Describes the deliverables acceptance criteria

    B. Describes how scope changes will be handled

    C. Describes the procedures for preparing the scope statement

    D. Describes the procedures for preparing the WBS

12. This involves changing the project or product scope without considering the impacts it will have to the project schedule, budget, and resources.

    A. Change request

    B. Scope creep

    C. Stakeholder/management directive

    D. Quality deficiency

13. There are three primary constraints on most all projects. Your customer, or project sponsor, will stipulate which of the three is the most important to them. Which three are the typical constraints found on the majority of projects? Choose three.

    A. Budget

    B. Team members

    C. Scope

    D. Quality

    E. Time

    F. Sponsors and stakeholders

    G. Scope management plan

14. Your project is underway, and a key stakeholder has submitted a change request. After further investigation, you discover some scope creep has also occurred on the project. These are examples of which of the following?

    A. Constraints

    B. Assumptions

    C. Influences

    D. Dependencies

15. Which term describes a characteristic of the scope planning (and other Planning) processes?

    A. Looping

    B. Iterative

    C. Ongoing

    D. Repetitive

16. Your project is nearing completion of the first phase. Your key stakeholder for this phase reminds you that she will not accept the deliverable unless it measures 3 centimeters exactly. If the measurements are off, phase 2 will be delayed, and the entire project will be at risk. This is an example of which of the following?

    A. Acceptance criteria

    B. Change request

    C. Stability of scope

    D. Product scope description

17. Elements of the project that are not listed on the WBS are considered what? Choose two.

    A. Work that will be completed in a future phase of the project

    B. Exclusions from scope

    C. They are considered scope creep.

    D. They are not considered part of the project.

    E. They are considered change requests.

18. You're developing a scope statement for a customer request. A couple of the elements that the customer wants could be difficult to accomplish, but after consulting with the project team, you think they can be done. These elements are not included in the product description. What should you do?

    A. Include these elements in the scope document, trusting your project team to deliver.

    B. Include these elements in the scope document, denoting them as a concern, and document how the issues were resolved.

    C. Discuss the problem elements with the project sponsor and the customer. Obtain sponsor sign-off.

    D. Note the elements in the exclusions section of the scope statement, and state that they'll be included in phase 2.

19. As you decompose the WBS, you discover a requirement that wasn't listed in the requirements section of the scope statement. This requirement is essential to a successful project because it details industry standard measurements that your team must comply with in order to complete the deliverable. Which of the following is true regarding this scenario?

    A. This is an example of a constraint.

    B. This is an example of an influence.

    C. This is an example of an assumption.

    D. This is an example of a dependency.

20. You're developing the scope statement for a new project. What project phase are you in?

    A. Initiating

    B. Planning

    C. Executing

    D. Controlling

Chapter 5

Creating the Project Schedule

THE COMPTIA PROJECT+ EXAM TOPICS COVERED IN THIS CHAPTER INCLUDE:

✓ **1.6 Given a scenario, execute and develop project schedules.**

- Scheduling activities
    - Determine tasks.
    - Determine task start/finish dates.
    - Determine activity/task durations.
    - Determine milestones.
    - Set predecessors.
    - Set dependencies.
    - Sequence tasks.
    - Prioritize tasks.
    - Determine critical path.
    - Allocate resources.
    - Set baseline.
    - Set quality gates.
    - Set governance gates.
        - Client sign off
        - Management approval
        - Legislative approval

✓ **4.1 Compare and contrast various project management tools.**

- Project Scheduling Software
- Charts
    - Gantt chart

Schedule Planning

After the WBS is completed, the next planning document you'll develop is the project schedule. As with the WBS, there isn't one right way to create the schedule. Here I've outlined the steps typically followed in creating a project schedule. On small projects you'll usually find all of these steps performed in one sitting, whereas larger projects may require separate sessions for each step in order to complete the schedule, and you may also need to perform some planning steps more than once as you gather information from stakeholders and team members. Here is the list of steps I typically follow when constructing a project schedule.

1. Determine tasks.

2. Sequence tasks.

3. Allocate resources.

4. Determine task durations including start and end dates.

5. Determine milestones.

6. Construct the schedule.

7. Determine the critical path.

8. Set the baseline and obtain approval.

9. Set quality gates.

10. Establish the governance process.

For the purposes of this exam, the terms *task* and *activity* are interchangeable.

At first glance, it would seem that putting together a schedule is fairly basic. All you need to do is enter the work packages from the WBS into a project-scheduling software program, and you have the schedule. However, a sound project schedule takes a lot of planning. All the tasks must be identified, they must be sequenced in the order they can be completed, they need an estimated time frame and effort for completion, resource assignments must be derived, and finally all this information must be organized logically to come up with the overall project schedule.

The schedule documents the planned start and finish dates of each of the tasks included in the project, and the total project duration is calculated once the schedule is complete. Once it is finalized, checking the schedule becomes part of the project manager's weekly, if not daily, routine until the project is completed. Progress is reported against the schedule, and status updates regarding activities are provided to the stakeholders on a regular basis.

If you don't take the time up front to create an accurate schedule, you'll be spending a lot of time during the project making changes to the schedule and explaining why deliverables are not being completed as anticipated.

You'll want several subject-matter experts to assist you with identifying and estimating tasks and creating the schedule, and it's a good idea to let everyone know up front that it could take more than one session to finalize the schedule.

Defining Tasks

The foundation for developing a project schedule is defining the list of tasks required to complete the project deliverables. This is an iterative process that involves further decomposing the WBS work packages into individual tasks. On small projects, it's a natural progression to break down the work packages into activities as you're decomposing the WBS because you have all the right people in the room and you've got the momentum going. Remember that tasks are not recorded on the WBS, but they are listed on the project schedule.

NOTE Keep tasks at a high enough level that they can be managed effectively without breaking them down so far that you're finding yourself managing each team member's to-do list.

It is helpful to have the team members who are assigned to the work package levels involved in task definition. Their expertise can help you not only with defining tasks but also with assigning resources and determining estimates for the tasks. A good rule of thumb I use for most projects is to define tasks at a level that will take 40 to 80 hours to complete. If you have a critical task that's shorter in duration than this, or a very small project, you may want to make an exception to this.

You'll want to list each activity you've defined on an *activity list* or *task list*. This list should include every activity needed to complete the work of the project, along with an identifier or code so that you can track each task independently. It's also good practice to list the WBS code this activity is associated with, along with a short description of the work.

Once you have all your tasks defined, you're ready to start putting them into the sequence in which they will be worked.

Task Sequencing

Sequencing is the process of identifying dependency relationships between project activities and sequencing them in proper order. First you need to identify the type of dependency, and then you need to determine the specific relationship between the activities.

Dependencies are relationships between activities. For example, one activity may not be able to start until another has finished, or perhaps one activity is dependent on another activity starting before it can finish. Let's look at several types of dependencies next.

Types of Dependencies

There are four categories of dependencies:

- Mandatory dependencies
- Discretionary dependencies
- External dependencies
- Internal dependencies

A *mandatory dependency* is directly related to the type of work being performed. For example, a utility crew can't lay the cable for a new housing area until a trench has been dug.

A *discretionary dependency* is defined by the project management team and is usually process or procedure driven and may include best-practice techniques. An example is a process that requires approvals and sign-off on planning documents before proceeding with the work of the project.

An *external dependency* is a relationship between a project task and some factor outside the project that drives the scheduling of that task. For example, installation of a new server depends on when the vendor can deliver the equipment.

An *internal dependency* is a relationship between tasks within an individual project and is therefore under the control of the project team.

It is important to know the type of dependency you're dealing with because you may have more flexibility with a discretionary dependency than a mandatory dependency. This distinction becomes important later when you're looking at ways to shorten the schedule and complete a project in less time.

Logical Relationships

It isn't enough just to know there is a dependency between two activities. You need to answer several other questions: How does the dependency impact the start and finish of each of the activities? Does one activity have to start first? Can you start the second activity before the first activity is finished? All these variables impact what your overall project schedule looks like.

Once you identify a dependency between two activities, you need to determine what that logical relationship is so that you can sequence the activities properly. Before covering those relationships, I'll present a few key terms related to understanding task dependencies.

A *predecessor* activity is one that comes before another activity. A *successor* activity is one that comes after the activity in question. Figure 5.1 shows a simple predecessor/successor relationship between Activity A and Activity B.

FIGURE 5.1   A predecessor/successor relationship

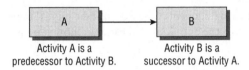

Activity A is a
predecessor to Activity B.   Activity B is a
                                    successor to Activity A.

Four possible *logical relationships* can exist between the predecessor activity and the successor activity. Understanding these relationships will help you determine if you can schedule the activities in parallel, or if one activity must wait until the predecessor task is completed. The four logical relationships are as follows:

Finish-to-start (FS)   In a finish-to-start relationship, the successor activity cannot begin until the predecessor activity has completed. This is the most frequently used logical relationship and is the default setting for most project-scheduling software packages.

Start-to-finish (SF)   In a start-to-finish relationship, the predecessor activity must start before the successor activity can finish. This relationship is seldom used.

Finish-to-finish (FF)   A finish-to-finish relationship is where the predecessor activity must finish before the successor activity finishes.

Start-to-start (SS)   In a start-to-start relationship, the predecessor activity must start before the successor activity can start.

Once the activity dependency relationships have been identified, your next step is creating a network diagram.

Creating a Network Diagram

One technique used by project managers to sequence activities is a network diagram. Understanding activity relationships is fundamental to using this technique. A *network diagram* depicts the project activities and the interrelationships among these activities. A network diagram is a great tool to develop with the project team. Use a whiteboard and label one sticky note with one activity. This will make it easy to see the workflow, and you can move the sticky notes around to make changes.

The most commonly used network diagramming method is the *precedence diagramming method (PDM)*. PDM uses boxes to represent the project activities and arrows to connect the boxes and show the dependencies. Figure 5.2 shows a simple PDM network diagram of tasks with finish-to-start dependencies and task durations.

FIGURE 5.2   The precedence diagramming method

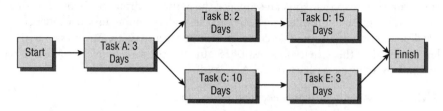

Now that the activities are sequenced based on their logical dependencies, you're ready to assign resources to the activities and estimate how long it will take to complete each activity.

Assigning Resources

Resources on a project schedule typically refer to human resources and/or consulting or contracting resources who will work on the tasks. As I mentioned earlier, it's a good idea to have the work package–level owners present while constructing the schedule, because they are typically the supervisors of the teams who will work on the tasks. They should know at a glance which person on their team to assign to the tasks. They are also typically expert enough in their areas to determine estimates and start and end dates for tasks. When there are multiple tasks within a work package, it's best to have the resources who are assigned to those tasks determine duration estimates.

Another helpful tool you can use when assigning resources is a resource calendar. The *resource calendar* describes the time frames in which resources are available. It defines a particular resource or groups of resources and may also include their skills, abilities, and quantity, as well as availability. Perhaps your project calls for a marketing resource and the person assigned to the marketing activities is on an extended vacation in October. The resource calendar would show this person's vacation schedule. Resource calendars also examine the quantity, capability, and availability of equipment and material resources that have a potential to impact the project schedule. For example, suppose your project calls for a hydraulic drill and your organization owns only one. The resource calendar will tell you whether it's scheduled for another job at the same time it's needed for your project.

Determining Task Durations

Determining task durations is the next step in constructing the project schedule. Duration estimating can be as easy as an expert giving you an educated estimate based on their experience, or it can be a complex process involving techniques and calculations to develop estimates—albeit most of these estimates are still based on expert opinions.

Before explaining some of the techniques you can use to complete your task duration estimates, let's make sure we have a common understanding of activity duration.

Defining Duration

When you are estimating duration, you need to make sure that you are looking at the total elapsed time to complete the activity. For example, let's say you have a task that is estimated to take five days to complete based on an eight-hour workday. You have one full-time resource assigned to this task, but they have only four hours a day to work on it. That means the actual duration estimate for this task is 10 days.

You also need to be aware of the difference between workdays and calendar days. If your workweek is Monday through Friday and you have a four-day task starting on Thursday, the duration for that task will be six calendar days because no work will be done on Saturday and Sunday. Figure 5.3 illustrates this situation. The same concept applies to holidays or vacation time.

FIGURE 5.3   A four-day task separated by the weekend

6 Calendar Days, 4 Workdays

Thursday	Friday	Saturday, No Work	Sunday, No Work	Monday	Tuesday

Make certain that everyone who is providing estimates is in agreement up front as to whether they will be provided in workdays or calendar days. We recommend using workday duration estimates. Most project management software packages allow you to establish a calendar that accounts for non-workdays and will exclude these days when computing duration.

Now that we have a common understanding of duration, I'll discuss the different techniques used to create duration estimates.

Estimating Techniques

You can use several techniques to determine task duration estimates. You'll look at three of the most common methods.

Analogous Estimating   *Analogous estimating* also known as *top-down estimating* is a technique that uses actual durations from similar tasks on a previous project. This is most frequently used at the early stages of project planning, when you have limited information about the project. Although analogous estimating can provide a good approximation of task duration, it is typically the least accurate means of obtaining an estimate. No two projects are the same, and there is the risk that the project used to obtain the analogous estimates is not as similar to the current project as it appears.

Results from analogous estimating are more accurate if the person doing the estimating is familiar with both projects and is able to understand the differences that could impact the activity durations on the new project.

Expert Judgment *Expert judgment* is a technique where the people most familiar with the work determine the estimate. Ideally, the project team member who will perform the task should provide the estimate. If all the team members haven't been identified yet, recruit people with expertise for the tasks you need estimated. Ask for people who have completed a similar task on a previous project to assist with the estimates for this project.

Remember that people with more experience will likely provide a shorter estimate for an activity than someone who doesn't have as much experience. You should validate the estimate or ask other experts in the department to validate it for you.

Parametric Estimating *Parametric estimating* is a quantitatively based estimating method that multiplies the quantity of work by the rate. To apply quantitatively based durations, you must know the productivity rate of the resource performing the task or have a company or industry standard that can be applied to the task in question. The duration is obtained by multiplying the unit of work produced by the productivity rate. For example, if a typical cable crew can bury 5 miles of cable in a day, it should take 10 days to bury 50 miles of cable.

PERT The *Program Evaluation and Review Technique (PERT)* is a method that the U.S. Navy developed in the 1950s. The Navy was working on one of the most complex engineering projects in history at the time—the Polaris Missile Program—and needed a way to manage the project and forecast the project schedule with a high degree of reliability. PERT was developed to do just that.

PERT and three-point estimates are similar techniques. The difference is that three-point estimates use an average estimate to determine project duration, while PERT uses what's called *expected value* (or the weighted average). Expected value is calculated using the three-point estimates for activity duration and then finding the weighted average of those estimates.

The formula to calculate expected value is as follows:

(optimistic + pessimistic + (4 × most likely)) / 6

Using the same numbers used in the three-point estimates produces the following expected value:

(7 + 14 + (4 × 10)) / 6 = 10 days

A PERT chart is another form of network diagram. The nodes (usually rectangles) represent milestones, the lines represent the sequence of tasks, the numbers near the lines represent the duration of the tasks. Figure 5.4 shows a simple PERT chart.

FIGURE 5.4 A sample PERT chart

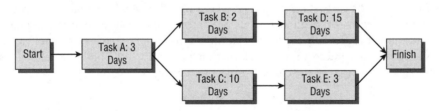

Keep in mind that most project managers use some combination of estimating techniques to determine task durations.

NOTE   You will not likely encounter questions on the exam regarding how to calculate three-point estimates or PERT estimates, but there could be questions about the techniques in general.

Now that you know their durations and dependencies, you can establish start and end dates for each task and construct the project schedule. You'll look at that next.

 Real World Scenario

The Bathroom Remodel Project

You have been assigned to manage the bathroom remodel project for your building. All restrooms will undergo a remodel, including the two large restrooms in the basement that also have lockers and showers.

Team members from the facilities department will be working on this project along with a contractor who was hired to perform the plumbing work. Some of the major deliverables include removing all the old fixtures and stalls, removing the existing tile, reconfiguring the space to add additional wheelchair-accessible stalls in each restroom, removing the old lockers, and installing all-new fixtures, stalls, counters, sinks, showers, and lockers.

You work with the team members from the facilities department to break down the tasks from each work package level. They inform you of predecessor activities, such as running electric lines for overhead lighting before setting countertops or mirrors. Because they have been involved in other projects of similar nature, they use expert judgment to provide you with duration estimates for their tasks and provide a rough order-of-magnitude estimate for the plumbing work. You all agree that the work should be estimated in workday increments.

Once all the work of the project is recorded on the schedule, you assign resources and calculate the duration of all the critical path tasks, and you find that the project will take 120 days to complete.

Creating the Project Schedule

Creating the schedule involves all the work you've done so far, including defining the tasks, sequencing the tasks, and determining duration estimates. You will now plug this information into the schedule and establish a start date and a finish date for each of the project activities. Let's walk through an example. Your project involves painting a house. One of the tasks is scraping the old paint off the walls. Another task is applying the new paint to the walls, and another task is painting the trim. You really shouldn't apply the new paint until the scraping is finished. That means the scraping task is a predecessor to the painting task. And the painting can't start until the scraping is finished (an FS relationship). You know from the experts providing estimates for these tasks that the scraping task will take two days and painting will take five days. If scraping starts on Wednesday, June 1, scraping should end on Thursday, June 2. All day June 1 and all day June 2 are spent on scraping. That means painting can start on Friday, June 3. The painters work on Saturday but not Sunday. If this is a five-day task, painting will finish on Wednesday, June 8. The total duration for these two tasks is eight days (including the Sunday the painters don't work).

It may take several iterations to get the schedule finalized. Once it's approved, it serves as the schedule baseline for the project. Once you begin the work of the project, you'll use this baseline to track actual progress against what was planned.

Before creating the schedule, let's revisit milestones and how they interact with the project schedule.

Milestones

Milestones are typically major accomplishments of the project and mark the completion of major deliverables or some other key event in the project. For example, approval and sign-off on project deliverables might be considered milestones. Other examples might be the completion of a prototype, functional testing, contract approval, and so on. A milestone is typically denoted on a project schedule as an event that is achieved once all the deliverables associated with that milestone are completed and it has a duration of zero.

Milestone charts are one method to display your schedule information. A milestone chart tracks the scheduled dates and actual completion dates for the major milestones. Table 5.1 shows an example milestone chart. As the project manager, you should pay close attention to milestone dates because they are also a communication trigger. Stakeholders need to be informed when major deliverables are completed or when a project has successfully moved to a new phase. If these dates are not met, you need to communicate the current status, the plans to bring the project back on track, and the new milestone date.

TABLE 5.1 A sample milestone chart

Milestone	Scheduled Start Date	Actual Start Date	Scheduled Completion Date	Actual Completion Date
Sign-off on scope statement	12/18	12/18	12/18	12/18
Sign-off on contract	2/02	2/02	2/02	2/02
Acceptance of deliverable 1	3/05	3/05	5/31	6/07
Acceptance of deliverable 2	3/15	4/01	6/30	7/15
Testing completed	7/01	7/16	7/31	8/16
Project acceptance and sign-off	8/10	8/20	8/10	8/20

Displaying the Schedule

Project management software is a tool that can save you a lot of time in creating your schedule. You can enter tasks, durations, and/or start and end dates; assign resources; and generate a graphical representation of the project. The most common way to display project schedules is a *Gantt chart.*

Gantt charts can show milestones, deliverables, and all the activities of the project including their durations, start and end dates, and the resources assigned to the task. Gantt charts typically display the tasks using a horizontal bar chart format across a timeline. I know project managers who have constructed Gantt charts for small projects using only a spreadsheet. Figure 5.5 shows a sample Gantt chart.

FIGURE 5.5   Gantt chart

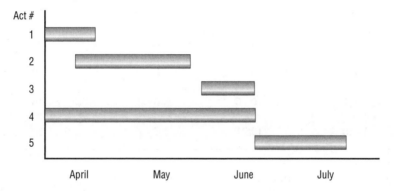

Network diagrams can also be used to display the project schedule if you add dates and durations in the boxes on the diagram.

The Critical Path Method

One of the most widely used techniques in schedule development is the *critical path method (CPM).* CPM determines the amount of *float time* for each activity on the schedule by calculating the earliest start date, earliest finish date, latest start date, and latest finish date for each task. Float is the amount of time you can delay the earliest start of an activity without delaying the ending of the project. Tasks with the same early and late start dates and the same early and late finish dates have zero float and are considered critical path tasks. If a critical path task does not finish as scheduled, the project end date will be affected.

For the exam, remember that tasks with zero float are critical path tasks. It is possible to calculate float time manually, but the formulas are beyond the scope of this book and the exam. Float time calculations are easily performed in project-scheduling software.

The *critical path* is the longest full path on the project. If you refer to Figure 5.2, you'll see that the longest path for this project has a duration of 20 days. This is calculated by adding the durations for the A-B-D path (Task A, Task B, and Task D). Simply add up the task durations for that path.

Table 5.2 shows the start and end dates and durations for the project tasks in table form. When calculating the start and end dates, the first day counts as day 1. Task A starts on 10/6, and it's worked on all day on the 6th, 7th, and 8th because it has a duration of three days. The next task can start on the 9th. In this example, there are no holidays or vacation days.

TABLE 5.2   Finding a critical path

Task	Start Date	End Date	Predecessor	Duration in Days
A	10/6	10/8	None	3
B	10/9	10/10	A	2
C	10/9	10/18	A	10
D	10/11	10/25	B	15
E	10/19	10/21	C	3

Again refer to Figure 5.2 and Table 5.2. What happens to the critical path if you eliminate Task E? Nothing changes, because the longest path (or the activities with the longest durations that are in the same path or have the same predecessors) is still A-B-D with 20 days. Path A-C-E has a duration of 16 days. If you eliminated Task B, the critical path changes to 18 days, which is along the new path A-D.

For the exam, make certain you understand that the critical path is the longest full path on the project. The simplest calculation you can use for the exam is to add up the duration of each activity for each path on the project and determine which one is the longest. It might be helpful to draw a network diagram on your scrap paper so that you can more easily see which tasks are dependent on each other and add up the durations of each path.

In addition to calculating the overall time to complete the project and identifying tasks on the critical path, CPM provides other useful information. You will be able to determine which tasks can start late or can take more time than planned without impacting the project end date. During project execution, the project manager can use this information to focus attention on the tasks that have the most impact on the overall project completion date.

There are times when you complete the schedule, calculate the critical path, and find that the duration of the project is unacceptable to the project stakeholders. If you find yourself in that situation, you can use duration compression techniques to help shorten the schedule. You'll learn about those techniques next.

Duration Compression

What happens if your calculation of the total project duration is longer than your target project completion date?

This is where *duration compression* scheduling techniques come into play. These techniques can be used during planning to shorten the planned duration of the project or during project execution to help resolve schedule slippage. The two duration compression techniques are crashing and fast tracking. You'll learn about both next.

Crashing

Crashing is a technique that looks at cost and schedule trade-offs. Crashing is typically implemented by adding more resources to the critical path tasks in order to complete the project more quickly. Crashing can also be accomplished by requiring mandatory overtime for those team members working on critical path tasks or by speeding up delivery times from team members, vendors, and so on.

One common misconception about adding resources is that if you double the resources, you can cut the duration in half. In other words, if two people can finish the work in four weeks, then four people must be able to finish in two weeks. This isn't always the case. Typically, the original resources assigned to the task are less productive when you add new resources because they're busy helping the new resources come up to speed on the work. Or, you may have so many resources working on the project that they are in each other's way.

Crashing can produce the desired results if used wisely, but you should be aware that crashing the schedule may increase risks and will impact your budget. Be certain you've examined these impacts to the project when using this technique.

Fast Tracking

Fast tracking is performing multiple tasks in parallel that were previously scheduled to start sequentially.

Let's go back to our painting example. Scraping and painting cannot start at the same time. However, we have two painting tasks. One is painting the walls; the other is painting the trim. The schedule currently shows the trim starting when the paint finishes (an FS relationship). We could fast-track these activities and have the trim start at the same time as the walls. Logistically, the crews will have to start from different points in the building so they aren't in each other's way, but the tasks can be started in parallel. There is a great deal of risk in fast tracking. If you decide to compress your project schedule using this method, be sure you get input from the team members as to what could go wrong. Document all the risks, and present them to your sponsor and other key stakeholders. Don't make the mistake of trying to do the project faster without communicating any of the potential risks or impacts. You need to make sure that everyone understands the potential consequences and agrees to the schedule change.

Project-Scheduling Software

Project-scheduling software is a tool that can save you a lot of time. It will automatically calculate task durations, determine the critical path, and help you balance resources. It provides you with the ability to display a number of different views of the project, which can be a great communication tool, and you can tailor the views for your audience. For example, executives typically don't want to see every line in a project schedule. They are interested in milestone views or in seeing the major deliverables and their due dates. The software will also allow you to save a baseline schedule for historical reference. Let's look at that next.

Setting the Baseline and Obtaining Approval

The *schedule baseline* is the final, approved version of the project schedule that includes the baseline start and finish dates and resource assignments. It's important that you obtain sign-off on the project schedule from your stakeholders and the functional managers who are supplying resources to the project. This ensures they have read the schedule, understand the dates, and understand the resource commitments. Ideally, it will also keep them from reneging on commitments they've made and promises of resources at specific times on the project.

The schedule baseline will be used throughout the project to monitor progress. This doesn't mean that changes can never be made to the schedule. But we encourage you to consider seriously any changes that will change the project end date. This will require modifying the schedule baseline and obtaining approvals through a change management process, which you'll learn about in Chapter 9.

Quality Gates

Quality gates in the project schedule are similar to milestones. They don't produce something per se; they are used to determine quality checks at strategic points in the project and ensure that the work is accurate and meets quality standards. In the painting example, you may have a quality gate that occurs after the scraping is finished but before the painting task begins. Painting should not begin until the quality gate is verified. In this case, the quality gate is to assure the work has been performed correctly and completely.

Your organization or PMO may have processes, checklists, or templates for use during quality gates. The activities associated with quality gates are not usually unique to the project; rather, they pertain to the product or service and can be used repeatedly on multiple projects.

Establishing Governance Gates

Governance gates are used as approval points in the project. On large projects, they can also be used as additional approval checkpoints or go/no-go decision points during the project.

The CompTIA objectives list three governance gates: client sign-off, management approval, and legislative approval. You learned about sponsor and stakeholder sign-offs earlier. You'll also want your customer to sign off on the project schedule (and the project plan when completed) as well. Remember that sign-offs help assure adherence to the schedule, agreement to the dates, and agreement to resource commitments. Legislative approvals, in my experience, actually occur prior to starting the project. However, I can imagine that a really large, complex public-sector project, such as building a space station, may require periodic legislative approval and oversight at strategic points in the project.

 Real World Scenario

Main Street Office Move: Scheduling

One of the deliverables noted in the scope statement is "Survey the Main Street Office Building and work with functional managers to determine seating charts." Working with two key functional managers on the project, you define the following tasks:

- Obtain floor plans for the Main Street Office Building.

- Perform a walk-through of the Main Street Office Building.

- Interview functional managers.

- Obtain current organizational chart from each functional manager.

- Provide draft seating charts to functional managers.

- Determine office placements and sizes.

- Determine cubicle placements.

- Determine printer placement.

- Determine breakroom areas.

- Provide final seating charts to functional managers.

The next step involves sequencing these tasks in the correct order and determining start and finish dates. You determined the following dependencies:

Task Number	Task	Dependency
1	Obtain floor plans.	NA
2	Building walk-through	NA
3	Interview functional managers.	Task 1
4	Obtain current org charts.	Task 3
5	Office and conference room placements and sizing	Tasks 1–3
6	Determine cubicle placements.	Tasks 1–3 and "Interior Designer" deliverable
7	Determine office placements.	Tasks 1–3
8	Determine printer locations.	Tasks 1–2
9	Determine breakroom areas.	Tasks 1, 2 and "Interior Designer" deliverable
10	Seating chart drafts to managers.	Tasks 1–9
11	Functional managers review seating chart drafts.	Task 10
12	Final seating chart approved.	Task 11

You need to determine some rough order-of-magnitude estimates for these tasks to determine start and end dates. For example, based on the recommendation from the functional managers that the final seating chart be approved before the cube buildouts begin, you know that Task 12 must finish before you place the order for the cubicle furnishings.

When breaking down the tasks for the "Procure and install cubicle furnishings" deliverable and determining start and finish dates, you realize you are working with a constraint that requires all cubicles and office furniture be delivered and installed by December 31. When you met with the IT director shortly after the scope statement was written, you discovered two weeks are needed to place computer equipment, monitors, and telephones on each desk. The cubicle buildouts and furniture delivery are predecessors to the computer equipment setup deliverable, so you need to determine when the cubicle furnishing order should be placed.

You've obtained an estimate from a vendor that it takes roughly two days to install 250 cubicles. Based on assumptions you've made in your conversations with Kate and the two functional managers who helped develop the task list, you think there will be a minimum of 1,000 cubicles. To allow IT two weeks to install the computer equipment, you determine delivery of the cubicle materials must occur by December 6. Cubicle buildouts can begin December 7 and must finish by December 15 (taking into account weekend days).

In summary, your cubicle furnishing order will take six weeks to fill according to your vendor contacts. If the materials must be delivered no later than December 6, that means Task 12 from the "Survey the Main Street Office Building" deliverable has to finish by October 25. You can work backward from October 25 to determine the start and end dates for each task in this deliverable.

You continue breaking down each deliverable into a list of tasks, sequence them in the proper order, determine start and end dates, and assign resources to the tasks to come up with the final schedule.

On the project schedule, you also add two quality gates: one after the general contractor has completed the floor remodels and one after the cubicles are built.

The governance gates include management sign-off on the building remodel plans, cubicle and office placements, and final approval when the move is completed.

Once the schedule is approved by the management team, it becomes the baseline schedule for the project, and all changes to the schedule must follow the change management process.

Summary

Many steps are involved in schedule planning. Task definition takes the work packages from your WBS and breaks them down into individual tasks that can be estimated and assigned to team members. Sequencing looks at dependencies between tasks. These dependencies can be mandatory, discretionary, internal, or external. A dependent task is either a successor or a predecessor of a linked task.

There are four types of logical relationships: finish-to-start, start-to-start, start-to-finish, and finish-to-finish. Duration estimating is obtained using analogous (also called top-down) estimating, parametric estimating, and expert judgment.

The critical path method (CPM) creates a schedule by determining float time. Float is the difference between the early and late start dates and the early and late finish dates. The critical path is the longest full path on the project.

Duration compression is the technique used to shorten a project schedule to meet a mandated completion date. Crashing shortens task duration by adding more resources to the project. Fast tracking is where two tasks are started in parallel that were previously scheduled to start sequentially.

A project schedule may be displayed as a milestone chart. Milestones mark major project events such as the completion of a key deliverable or project phase. Gantt charts are a common method to display schedule data as well. The completed, approved project schedule becomes the baseline for tracking and reporting project progress.

Quality gates may be added to the schedule to determine whether the work so far is accurate and meets quality standards. Governance gates include client sign-off, management approval, and legislative approval.

Exam Essentials

Describe the sequencing process.   Sequencing is the process of identifying dependency relationships between the project activities and scheduling activities in the proper order.

Name the two major relationships between dependent tasks.   A predecessor is a task that exists on a path with another task and occurs before the task in question. A successor is a task that exists on a common path with another task and occurs after the task in question.

Name the four types of logical relationships.   The four types of logical relationships are finish-to-start, start-to-start, start-to-finish, and finish-to-finish.

Know and understand the three most commonly used techniques to estimate activity duration.   Expert judgment relies on the knowledge of someone familiar with the tasks. Analogous or top-down estimating bases the estimate on similar activities from a previous project. Parametric estimates are quantitatively based estimates that typically calculate the rate times the quantity.

Define the purpose of CPM.   CPM calculates the longest full path in the project. This path controls the finish date of the project. Any delay to a critical path task will delay the completion date of the project.

Explain a network diagram.   A network diagram is used to depict project activities and the interrelationships and dependencies among these activities.

Name the three most common ways project schedules are displayed.   Project schedules are typically displayed as milestone charts, PERT network diagrams, or Gantt charts; a Gantt chart is a type of bar chart.

Define quality gates and governance gates.   Quality gates are used to check the work, and governance gates are used as client sign-offs, management approvals, and legislative approvals.

Key Terms

Before you take the exam, be certain you are familiar with the following terms:

activity list

analogous estimating

crashing

critical path

critical path method (CPM)

dependencies

discretionary dependency

duration compression

expert judgment

external dependency

fast tracking

float time

Gantt chart

governance gates

internal dependency

logical relationship

mandatory dependency

network diagram

parametric estimating

precedence diagramming method (PDM)

predecessor

Program Evaluation and Review Technique (PERT)

quality gates

resource calendar

schedule baseline

sequencing

successor

task list

top-down estimating

Review Questions

1. Which of the following is not true for the critical path?

   A. It has zero float.

   B. It's the shortest activity sequence in the network.

   C. You can determine which tasks can start late without impacting the project end date.

   D. It controls the project finish date.

2. You are a project manager for a major movie studio. You need to schedule a shoot in Denver during ski season. This is an example of which of the following?

   A. External dependency

   B. Finish-to-start relationship

   C. Mandatory dependency

   D. Discretionary dependency

3. What is analogous estimating also referred to as?

   A. Bottom-up estimating

   B. Expert judgment

   C. Parametric estimating

   D. Top-down estimating

4. You are working on your network diagram. Activity A is a predecessor to Activity B. Activity B cannot begin until Activity A is completed. What is this telling you?

   A. There is a mandatory dependency between Activity A and Activity B.

   B. There is a finish-to-start dependency relationship between Activity A and Activity B.

   C. Activity A and Activity B are both on the critical path.

   D. Activity B is a successor to multiple tasks.

5. What are the most commonly used forms to display project schedules? Choose two.

   A. PERT charts

   B. Gantt charts

   C. CPM diagrams

   D. A spreadsheet

6. What are the crashing and fast tracking techniques used for?

   A. Duration compression

   B. Activity sequencing

   C. Precedence diagramming

   D. Task definition

7. Which of the following is true for float?

   A. It's calculated by adding the durations of all tasks and dividing by the number of tasks.

   B. It's time that you add to the project schedule to provide a buffer or contingency.

   C. It's the amount of time an activity can be delayed without delaying the project completion date.

   D. It is calculated only on the longest path of the project schedule.

8. You have added a checkpoint in the project schedule to determine whether the work is accurate. What is this called?

   A. Governance gate

   B. Quality gate

   C. Approval gate

   D. Checkpoint gate

9. Activity B on your project schedule starts on Monday, October 3, and ends on Wednesday, October 12. You calculate duration in workdays, and the team does not work on Saturdays or Sundays. How many days is the total duration of this task in workdays?

   A. 9 days

   B. 10 days

   C. 8 days

   D. 7 days

10. Which of the following is not true for critical path tasks?

    A. The early start date is less than the late start date.

    B. These activities are on the longest full path on the project schedule.

    C. The float time for tasks is zero.

    D. The late finish date is the same as the early finish date.

11. Starting two activities at the same time that were previously scheduled to start sequentially is known as which of the following?

    A. Fast tracking

    B. Mandatory dependency

    C. Crashing

    D. Baselining

12. Your task requires 4 miles of paving, and it will take 30 hours to complete a mile. On a past project similar to this one, it took 150 hours to complete. Which of the following is true regarding this estimate?

    A. The total estimate for this task is 120 hours, which was derived using expert judgment.

    B. The total estimate for this task is 120 hours, which was derived using parametric estimating.

    C. The total estimate for this task is 150 hours, which was derived using analogous estimating.

    D. The total estimate for this task is 150 hours, which was derived using expert judgment.

13. Governance gates include all of the following except for which one?

    A. Assigning a go/no-go decision point

    B. Obtaining client sign-off

    C. Securing management approvals

    D. Obtaining legislative approvals

    E. Assuring the work is performed correctly

14. You're working on a project that you and your team have estimated takes 15 business days to complete. Your team does not work on Saturday, Sunday, or holidays. Given the work calendar shown here and a start date of Tuesday, November 1, what is the scheduled completion date for this task?

Sun	Mon	Tue	Wed	Thu	Fri	Sat
		1	2	3	4	5
6	7	8	9	10	11 Holiday	12
13	14	15	16	17	18	19
20	21	22	23	24 Holiday	25	26
27	28	29	30			

    A. 21

    B. 15

    C. 22

    D. 23

15. All of the following are true regarding milestone charts except for which one?

    A. Milestone charts list the major deliverables or phases of a project.

    B. Milestone charts show the scheduled completion dates.

    C. Milestone charts show the actual completion dates.

    D. Milestone charts are commonly displayed in bar chart format.

16. You're in the process of developing a project schedule for a new project. You've just completed the WBS. What would be the project manager's next step in figuring out what tasks go into the project schedule?

    A. Develop a task list.

    B. Determine the critical path tasks.

    C. Develop a project schedule.

    D. Estimate activity duration.

17. After the schedule is approved by the sponsor, customer, and key stakeholders, what happens next?

    A. The resources are assigned.

    B. The schedule baseline is established.

C. The task start and end dates are finalized.

D. The critical path is finalized.

18. Which is the most commonly used logical relationship?

A. Finish-to-start

B. Start-to-finish

C. Start-to-start

D. Finish-to-finish

19. How long is the critical path in days in the graphic shown here if you eliminate Task B?

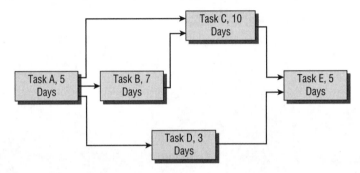

A. 13 days

B. 20 days

C. 27 days

D. 30 days

20. The following exhibit shows a series of tasks in a project schedule. Which path represents the critical path?

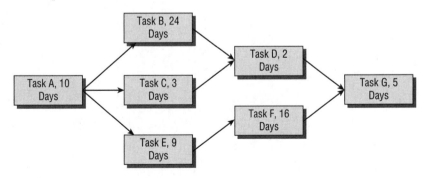

A. A-E-F-G

B. A-C-D-G

C. A-B-D-G

D. A-B-D-F-G

Chapter

6

Resource Planning and Management

THE COMPTIA PROJECT+ EXAM TOPICS COVERED IN THIS CHAPTER INCLUDE

✓ **1.8 Explain the importance of human resource, physical resource, and personnel management.**

- Resource management concepts
  - Shared resources
  - Dedicated resources
  - Resource allocation
  - Resource shortage
  - Resource overallocation
  - Low quality resources
  - Benched resources
  - Interproject dependencies
  - Interproject resource contention
- Personnel management
  - Team building
  - Trust building
  - Team selection
  - Skill sets
  - Remote vs. in-house
  - Personnel removal/replacement
  - Communication issues

- Conflict resolution
  - Smoothing
  - Forcing
  - Compromising
  - Confronting
  - Avoiding
  - Negotiating

✓ **4.1 Compare and contrast various project management tools.**

- Responsible, Accountable, Consulted, Informed (RACI) Matrix

Project execution is where the work of the project is performed, and that work is performed by people. Successful project execution involves developing the project team, managing the work according to the project plan, and managing conflict.

You'll have relationships with a number of individuals and groups during the life of the project. All of your people management skills will come into play as you negotiate with the sponsor, team members, vendors, functional managers, clients, users, and other internal organizations.

If you talk to veteran project managers about what makes their projects successful, most will list the project team. Understanding how to build this temporary group into a team, making sure appropriate training is provided, managing conflict, and implementing a meaningful rewards and recognition plan are all challenges you'll face in developing a cohesive team.

Along the way, you'll discover some bumps in the road that will require specific actions to get the project back on course. This may also require updating project plan documents. This chapter starts by determining resource needs. Let's dive in.

Determining Resource Needs

All projects require resources to complete the tasks and deliver a successful project. The CompTIA exam breaks the concept of resources into several categories: resource management concepts, personnel management, and conflict management. First, you'll start with resource management concepts and explore the idea of shared vs. dedicated resources, resource overallocation, and interproject dependencies. You'll look at personnel management and conflict management later in this chapter.

Resource Management Concepts

Let's focus on some of the characteristics of resources and some of the management concepts you might encounter on the exam.

First, remember that resources on a project do not always mean human resources. For example, equipment and materials are a type of *physical resource*. In Chapter 5 I talked about using a hydraulic drill on the project and using the resource calendar to determine the availability of this resource. I also talked previously about functional, matrix, and projectized organizations. In projectized organizations, you'll recall that resources are typically dedicated to the project. *Dedicated resources* are the ideal scenario for a project manager because you have full authority and control of the resource time and the tasks they work on. You don't have to coordinate schedules with another manager or fear having the resource pulled off the project because of an "emergency" in the functional area.

In a functional or matrix organization, you'll find you often have to share resources. A *shared resource* works for both the functional manager and the project manager. Typically, the team member will remain loyal to the person writing their performance appraisals and reviews. And in my experience, shared resources are often torn between their functional work and project work and find themselves defaulting to functional work because the business has to continue operating and they fear their work might pile up while they are busy working on project tasks. Project managers working with shared resources should negotiate with the functional managers to make certain the shared resource has the appropriate amount of time available to work on the project. They should also negotiate the ability to have some say in the team member's performance ratings. This will help you reinforce that the project work is important and needs to be addressed and prioritized by the team member.

Another element to keep in mind when working with functional managers is that they may jump eagerly at the opportunity to assign resources to your project. However, buyer beware! I have experienced that having overeager functional managers sometimes means the resources they're giving me are *low-quality resources*. This doesn't mean they're bad resources, but it generally means they don't have skill sets needed or, worse, they may have abrasive personalities or a history of conflicts with other team members. If you have no choice, I recommend meeting with such a team member as soon as possible and setting clear expectations. I'll talk more about this concept in the "Personnel Management" section of this chapter. Also, beware of low-quality physical resources that won't hold up to the wear and tear needed to complete the task.

A *resource shortage* can also lead to a *resource overallocation*. In other words, because of the shortage, the resource with the skills needed to perform your task is also assigned to other projects. Resource shortages are not always the cause of resource overallocation. It could be that you've scheduled the resource to complete more tasks than the time available. Overallocated resources will show up on a project schedule as 100 percent (or more) allocated. In reality, resources never have 100 percent of their time available for any task because they also have administrative functions they need to perform, such as filling out time cards, attending staff meetings, and answering emails.

 Generally speaking, when your organization has only one person or one physical resource that can perform a task, you will be dealing with overallocation issues.

I'm sorry to say there isn't an easy answer for overallocated resources. This might mean you'll have to work with the senior management team to determine project priorities. If the one and only resource needed for two competing projects can't work on both projects, management will have to decide which one has priority.

The opposite of overallocating team members is having *benched resources*. Benched resources are those who have typically finished a project and are not yet assigned to a new project or have a time gap between the finish and start of the new project. Benched resources are costly to the organization because, well, they're being paid to sit around and wait for the next assignment. This scenario typically occurs in a projectized organization.

 **Real World Scenario**

The Equipment Was There, but the Electricity Wasn't!

Jill is a project manager for a large corporation based in the Pacific Northwest. The corporate managers decided to build a new building to house all the departments in one place. The company planned to save money by reducing lease contracts while increasing the level of efficiency because coworkers would be in closer proximity to one another. Jill was put in charge of a project to move all the people into the new building. The project would take about six months, and she would be required to move 1,000 people in "move waves," with a total of six waves.

Shortly after she received the project, Jill met with the building's contractor to discuss the location of the different departments and the power requirements and to determine lighting diagrams for the cubicles throughout the building and the data center. The contractor told Jill that, in the interest of saving money, the corporate engineers had opted to reduce the number of electrical cables in the data center, though he assured her he had planned for enough connections.

During the first-wave move, some, but not all, of the servers that were going into the data center were delivered, and the various administrators were there to hook them up. Jill was shocked to find out that all the electrical connections were used up in the first wave! Even though there were more servers to come, she had nowhere for them to connect to power. The assurance that the contractor gave her was suddenly out the window.

Jill went back and inventoried the electrical requirements for the remaining servers and discovered some had regular 15-amp requirements, others needed 20-amp circuits, and still others required a specialized 277/480-amp circuit. When she informed the contractor of this discovery, she found that he had only installed 15-amp circuits and wasn't aware of the other power requirements.

Jill assessed the number of circuits she needed for each of the remaining servers and their power requirements, and the total came to seventeen 15-amp, ten 20-amp, and two 277/480-amp circuits. Next she went back to the contractor to get an estimate of the cost for the 29 new circuits. That estimate came in at $17,500!

Finally, with much trepidation, she went to the project sponsor, explaining that she had overlooked the power requirements and that additional monies were required to complete the project.

In retrospect, Jill realized she should have used subject-matter experts to help determine the server and circuit requirements.

Interproject Work

Remember that when you're allocating resources you'll need to take *interproject dependencies* and *interproject resource contention* into consideration.

Interproject dependencies occur when you need the completed deliverables from one project in order to work on the current project. For example, perhaps your organization is constructing a new highway lane on an existing highway in your city. However, there are other projects associated with this one (part of the program) that must be completed before you can finish the highway lane construction. One of those projects is performing the environmental studies on the ground where the new lane is planned. Another might be digging a trench to lay fiber-optic cable along the highway. Both of these projects must be completed before you can begin your project. Thus, you have an interproject dependency.

It's important to meet with the project managers from the other projects and understand their schedules and risks. You should also discuss with them the possibility of removing dependencies wherever possible. Make certain that you are receiving up-to-date status reports on the other projects so that you can track the key dependencies and keep your project on schedule or take corrective action where necessary.

Interproject resource contention is similar in nature to the overallocation problem discussed previously. Your resource may be scheduled for similar tasks on other projects. You'll need to work in coordination with the project managers on those other projects to schedule the resource so that they are available to both of you when needed.

Personnel Management

Managing a project team differs from managing a functional work group. Project teams are temporary, and getting everyone to work together on a common goal can be challenging, especially if your team members are specialists in a given discipline and don't have a broad business background. As the project manager, you must mold this group into an efficient team that can work together to deliver the project on time, on budget, and within scope, all while producing quality results. This is not always an easy undertaking, especially if you factor in a combination of full- and part-time team members, technical and nontechnical people, resources from inside and outside the organization, and in some cases a team dispersed over a large geographic area.

Selecting Team Members

Choosing or acquiring team members with the right skills and demeanor is important to the success of your project. Project staff might come from inside (often referred to as *in-house resources*) or outside the organization. They may also come from *remote* locations. Resources located in other parts of the company, or other parts of the world, sometimes feel disconnected from the project team. When possible, you should *collocate* the resources; that is, you can bring them all together physically so they work in the same location.

You may often find that you don't have control over the selection of team members. Functional managers may assign team members to the project according to their availability, as we discussed earlier. Other times, you'll know the team members you want on the project and can request them. What's important is that when you're not familiar with the resource or their skill set, spend some time with the functional manager discussing the skills and abilities you need for the tasks. Interview the resource being recommended and ask specific skill-related questions and how they might go about approaching the tasks you'll be assigning to them. If they don't seem like a good fit or you are doubtful about their skills and abilities, meet with the functional manager, share your concerns, and ask them if there is someone else available.

Organization Charts and Position Descriptions

Once your team members have been selected, it's a good idea to create a *project organization chart*. We've all seen an organization chart. It usually documents your name, your position, your boss, your boss's boss, your boss's boss's boss, and so on. It's hierarchical in nature, similar to a WBS. In the case of a project organization chart, you can present the information in a couple of different ways, including the traditional org chart with the project manager and project team member names as shown in Figure 6.1.

FIGURE 6.1   Project organization chart

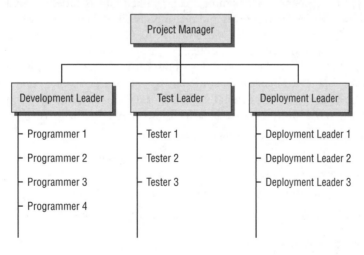

You may also present the work of the project in an *organization breakdown structure (OBS)*. This form of organization chart shows the departments, work units, or teams within an organization (rather than individuals) and their respective work packages.

A *resource breakdown structure (RBS)* is another type of hierarchical chart that breaks down the work of the project according to the types of resources needed. For example, you might have painters, carpenters, and electricians as resource types on the RBS.

Matrix-Based Charts

A *responsibility assignment matrix (RAM)* is a matrix-based chart that maps your WBS elements to the required resources. Table 6.1 shows a sample portion of a RAM for an IT development project. It lists the WBS identifier, the type of resource required, and the number of resources needed for each skill set.

TABLE 6.1   Sample project responsibility assignment matrix

WBS identifier	Programmer	Tester	Marketing	Tech writer	Server
10–1–1–1				1	
10–1–1–2	2				
10–1–1–3		3			1
10–1–1–4	4				
10–1–1–5			1		

Another type of responsibility assignment chart I like to use is called a RACI chart. These are used to show the types of resources and the responsibility they each have on the project. A RACI is usually depicted as a chart with resource names (or the individual names of team members) listed in each row and work elements such as milestones or work packages listed as the columns. Indicators in the intersections show where the resources are needed.

Table 6.2 shows a *RACI chart* for a conference event project team. In this example, the RACI chart shows the level of accountability each of the participants has on the project. The letters in the acronym RACI are the designations shown in the chart.

R = Responsible for performing the work

A = Accountable, the one who is responsible for producing the deliverable or work package and approves or signs off on the work

C = Consulted, someone who has input to the work or decisions

I = Informed, someone who must be informed of the decisions or results

TABLE 6.2   Sample RACI chart

	Olga	Rae	Charlie	Lolita
Hotel conference rooms	R	A	C	C/I
Food and snack service	C/I	C	I	A/R
Speaker contracts	C	I	R	A

*R = Responsible, A = Accountable, C = Consult, I = Inform

In this example, Olga is responsible for booking and assigning the hotel conference rooms, but Rae is accountable and is the one who must make sure this task is completed and approved. Charlie is responsible for securing speakers, but Lolita is accountable for making sure this task is completed. Charlie also needs to be consulted regarding hotel conference rooms so that speakers are assigned to a room large enough to hold the projected number of attendees. Lolita is responsible and accountable for the food and snack service and needs to be consulted and informed about the hotel conference room task.

This is a great tool because it shows at a glance not only where a resource is working but what the resource's responsibility level is on the project.

Roles and Responsibilities

A roles and responsibilities document lists each group or individual team member on the project and their responsibilities. But roles and responsibilities of project team members are more than just the assigned tasks. There are standards and methodologies to be adhered to, documentation to be completed, and time-reporting responsibilities, to name a few. So in addition to assigning people to tasks, it is a good idea to develop a template to document roles and responsibilities beyond just the task assignment. The more clarity around who is responsible for what, the better. We discussed the various roles and responsibilities of team members in Chapter 2 if you need a refresher on specific project roles. You could use that information as a starting point for the roles and responsibilities document.

Whatever format you choose to document the roles and responsibilities, the intent is to be as clear and precise as possible in defining the key areas of accountability for each team member.

Roles and responsibilities may change over the course of the project, so be sure to update this document as needed.

Building and Managing a Cohesive Team

Every team progresses through a series of development stages. It's important to understand these stages because team member behaviors will change as you progress through them, and the stage they're in affects their interactions with each other.

Dr. Bruce Tuckman developed a model that describes how teams develop and mature over time. According to Tuckman, all teams progress through the following five stages of development: forming, storming, norming, performing, and adjourning. You'll take a brief look at each of these next:

Forming   *Forming* is the beginning stage of team formation, when all the members are brought together, introduced, and told the objectives of the project. This is where team members learn why they're working together. During this stage, team members tend to be formal and reserved and take on an "all-business" approach.

Storming   *Storming* is where the action begins. Team members become confrontational with each other as they begin vying for position and control during this stage. They're working through who is going to have the most influence and they're jockeying for status.

Norming   *Norming* is where things begin to calm down. Team members know each other fairly well by now. They're comfortable with their positions in the team, and they begin to deal with project problems instead of people problems. Decisions are made jointly at this stage, and team members exhibit mutual respect and familiarity with one another.

Performing   Ahh, perfection. Well, almost, anyway. *Performing* is the stage where great teams end up. This is where the team is productive and effective. The level of trust among team members is high, and great things are achieved. This is the mature development stage. Not all teams make it to performing; many churn between storming and norming. It is a real joy to work with a team that has made it to the performing stage.

Adjourning   As the name implies, *adjourning* refers to breaking up the team after the work is completed and returning the team members to their functional managers and teams.

Different teams progress through the stages of development at different rates. When new team members are brought onto the team, the development stages start all over again. It doesn't matter where the team is in the first four phases of the development process—a new member will start the cycle all over again.

Progressing through these stages can be enhanced with the use of team-building activities. *Team building* is a set of activities or exercises designed to get a diverse group of people to work together in an efficient and effective manner. It helps them to form social bonds and identify common interests, and it can help clarify the roles and responsibilities of each team member. Organized team-building activities are most effective when a team is in the forming and storming stages, especially if they don't know each other well. Search the Web or ask your human resource department for team-building activities that will help your new staff get to know each other better. In my experience, once the team moves into the norming stage and beyond, there may not be as much of a need for organized team activities; however, informal activities are a great benefit in this stage, especially if the team arranges them, such as lunches, pizza nights, sports outings, beers after work, and so on.

You may also find team-building activities helpful if your team is experiencing personality clashes or if there are changes to the team makeup where old members have rolled off and new members have rolled on. Organizational changes are another good reason to employ team-building activities.

Trust Building

Building trust with and among your team members is, in my experience, the most important thing a project manager can do to ensure a top-performing team. Trust, as the old saying goes, is earned, not given.

To build and maintain the trust of your project team members, you need to demonstrate competence, respect, honesty, integrity, and openness. You must also demonstrate that you are willing to act on performance problems.

Trust building occurs over time, not overnight. In my experience, trust building includes doing what you say you'll do, supporting your team, showing concern for your team, having their back, putting the good of the organization above your own, and being humble. Not many folks want to work for or with someone whose primary concern is him- or herself.

Teams that trust one another and their project manager are more comfortable stating their opinions and objecting to ideas that don't make sense. This may sound contrary, but teams who are able to speak their minds are much more likely to be successful. They have buy-in to ideas and tasks because they had some say-so in the matter.

If the project manager is not open to this type of feedback or thinks they have all the answers and don't require input from the team, they aren't likely to experience their teams moving to the performing stage.

Take the time to get to know your team members. Ask them questions about their outside interests and show genuine concern when they have issues or conflicts. This can go a long way in establishing a trusting relationship.

Monitoring Team Performance

Managing team member performance can be a complex undertaking. A successful project manager understands that most people work at their best when they're allowed to do the work they were assigned without someone preapproving every action they take. As long as the end result is accomplished according to plan and there is no impact on scope, schedule, budget, or quality, team members should be given freedom and choices regarding how to complete their tasks.

Although you shouldn't micromanage team members, they do need feedback on how they're doing. Most team members perform well in some areas and need improvement in others. Even if your organization does not require project managers to conduct formal written appraisals, you should take the time to provide feedback to the team members and not get so caught up in managing the project issues that you neglect your team members. The following are important areas of focus as you prepare to discuss performance with your team members:

- Specifying performance expectations
- Identifying inadequate performance behaviors
- Rewarding superior performance
- Reprimanding inadequate performance
- Providing specific consequences for choices made

Performance Feedback

The first meeting you should have with any new team member, and at the beginning of every project, is the expectations-setting meeting. This is where you discuss the following: their role on the project, due dates, performance expectations, team interactions, industry standards and regulations that must be followed, and your expectations of their work quality. You'll want them to know the best way to contact you and that you are available for any questions they may have.

Performance feedback should be given in a timely fashion. It is of little value to attempt corrective action on something that happened several weeks or months ago. The team member may not even remember the specifics of the performance in question. It is best to deal with negative situations as soon as they occur. Be objective about your observations and ask the team member to explain the issue from their perspective. The best case is that you and the team member come to an understanding on what occurred and how the behavior should be corrected. The not-so-best case is you'll have to remove this team member.

You must work with your human resource department if the team member refuses to correct their behavior or if they don't have the ability to perform their tasks. If your team member came from a functional area, you'll need to write up the issues and provide this information to the functional manager.

It isn't fun to have to remove or replace personnel from a project. But if all other actions fail and the team member refuses to improve, you must take steps to prevent a serious decline in morale. Taking this action also builds trust in you as a leader. Your team will see that you're willing take the actions needed in a compassionate manner to deal with team issues.

Rewarding Superior Performance

Recognition and rewards are important elements of both individual and team motivation. They are formal ways of recognizing and promoting desirable behavior and are most effective when carried out by the project manager or management team. Project teams work hard and often overcome numerous challenges to deliver a project. If your company has a functional organizational structure, the project work may not receive the appropriate recognition from the functional managers. That means it's up to you to recognize the job your team is doing and implement a recognition and reward system.

When you think of rewards, you generally think of monetary rewards. And that's great if you're lucky enough to have money for a reward system, either as a direct budget line or as part of a managerial reserve. But there are options besides money that you can use as a reward—for example, time off, movie tickets, sporting or cultural events, team dinners, trophies, and so on. We've worked in organizations where an ordinary object was designated the "trophy" for outstanding performance. Individual team members were responsible for recognizing each other and passing on the trophy.

Another no-cost idea is a letter of recognition sent to an employee's manager, with copies to the appropriate organizational executives and the project sponsor. This can be a powerful means of communicating your appreciation for outstanding performance.

Not all project managers have the resources to reward team members either individually or collectively, but that does not mean superior performance should go unrecognized. One of the easiest things you can do is simply to tell people that you are aware of their accomplishments and that you appreciate their efforts. A simple thank-you works wonders. A handwritten note of appreciation only takes a few minutes and is something the employee can keep in their file.

Make certain that rewards are proportional to the achievement and are distributed equitably among team members. Playing favorites or recognizing the same team member with awards multiple times can kill morale rather than improve it.

Also, be sensitive to your team members and their cultural backgrounds. Some people love to be praised in public, and others may actually be shamed by this action rather than feeling appreciated.

You should develop and document the criteria for rewards, especially monetary awards. Work with your human resource department to understand the guidelines for reward systems in your organization.

The key is to establish a program to acknowledge the efforts of your project team members, whether it involves money, prizes, letters of commendation, or a simple thank-you. Whatever form your rewards and recognition program takes, you must make sure that it is applied consistently to all project team members and that the reward is appropriate for the level of effort expended or the results that were achieved. Inconsistent application of rewards is often construed as favoritism.

Rewards for superior performance can be given publicly, but a discussion of inadequate performance should always be done privately. Berating a team member in front of others is inappropriate and will likely make the person angry and defensive. It may also negatively impact the morale of the team members who witnessed the berating.

Developing your team and improving overall performance can also be accomplished through training. I'll discuss this next.

Training

Training involves determining your team member's skills and abilities, assessing the project needs, and providing the training necessary for the team members to perform their activities. In some industries or organizations, one of the perks associated with being assigned to a new project is the opportunity to expand a skill set or get training on new products or processes.

One of the most common types of training provided to project teams is project management training. Project management training may include a session developed by the project manager, formal training provided by an outside company, or training from an internal PMO on the standard methodologies, tools, and templates all project members are expected to use.

Conflict management is an important aspect of team building, communication, and team cohesiveness. We'll look at this topic next.

Real World Scenario

Project Management 101

One of the more successful experiences I've had with project management training involved a project team in an organization that was just starting to implement project management disciplines. Based on the chaos surrounding earlier attempts at running projects, it was clear that the team members needed a common understanding of what project management was all about.

We contracted with a professional project management training company to teach a beginning class in project management concepts. All project team members were required to attend this session. All of the exercises associated with the class were based on the actual project the team members were assigned to. Not only did the team members gain knowledge of the project management discipline, but they were able to contribute to the project itself while in class.

Although this took some time and money, it was well worth the effort. All the team members used common definitions of terms, and it was much easier to talk about the meeting requirements, the project baseline, scope creep, and other fundamental project management concepts. The success of this project resulted in the organization-setting goals regarding project management training for the entire department.

Conflict Management

One thing is certain: if you have people working on your project, you'll likely experience conflict at least once, if not many times, during the course of the project. *Conflict* is the incompatibility of desires, needs, or goals between two parties or individuals. This can lead to one party resisting or blocking the other party from attaining their goals.

Conflict may arise on a project for any of several reasons. As I've discussed in several places throughout the book, resources in most organizations are in high demand. Competition for resources can cause conflicts among the project managers, functional managers, and even project team members who may not be happy with less stellar selections as teammates.

Work styles can sometimes cause conflict. For example, we've all worked with team members whose desks were so buried in papers and books and other items that you couldn't see the desktop. And of course, we've seen the opposite as well—those team members without a speck of paper on the desk, only the telephone and a computer monitor.

Some team members are early birds and show up for work before the sun is up but are tired and cranky by 4 p.m., while others do their best work in the afternoon and early evening hours. There are hundreds of ways that work styles can vary and cause conflicts on the team. You should be aware of the preferences of your team members and accommodate reasonable solutions whenever possible.

Constraints are another area that can cause conflict on a project. Change requests, scope creep, and stakeholders are just a few examples of constraints that may drive incompatible goals.

Another common cause of conflict is communication issues. Perhaps team members don't understand the goals of the project, or maybe they lack solid interpersonal skills and are not adept at communicating their needs or issues. Sometimes team members don't communicate well with each other and deadlines are missed or tasks don't meet quality standards. Communication is such an important issue in project management I've dedicated a whole chapter to the topic. I'll talk more about communication in Chapter 8.

There are several techniques you can use to address and ideally resolve conflict among team members. You'll look at them next.

Managing Conflict

One of the most important concepts I can share with you that I've learned during my career managing hundreds of projects and personnel is that conflict will not go away on its own. You can't wish it away and hope for the best. Unfortunately, ignoring conflict will not make it go away either. I tried that tactic once or twice early in my career and it was a resounding failure. You need to address conflict head-on before it grows and gets out of hand.

 This is such an important concept that I'm going to say it again. As soon as you are alerted that conflict is lurking among the team, even if you simply suspect there is conflict (listen to your gut on this), deal with it immediately. Conflict is like a cancer and will grow and fester out of control without intervention.

According to CompTIA, there are several ways to detect and resolve conflict. You'll look at each technique next.

Smoothing   *Smoothing* is a temporary way to resolve conflict. In this technique, the areas of agreement are emphasized over the areas of difference, so the real issue stays buried. This technique does not lead to a permanent solution. Smoothing can also occur when someone attempts to make the conflict appear less important than it really is. Smoothing is an example of a lose-lose conflict-resolution technique because neither side wins.

Forcing   *Forcing* is just as it sounds. One person forces a solution on the other parties. Although this is a permanent solution, it isn't necessarily the best solution. People will go along with it because, well, they're forced to go along with it. It doesn't mean they agree with the solution. This isn't the best technique to use when you're trying to build a team.

This is an example of a win-lose conflict-resolution technique. The forcing party wins, and the losers are forced to go along with the decision.

Compromising   *Compromising* is achieved when each of the parties involved in the conflict gives up something to reach a solution. Everyone involved decides what they'll give on and what they won't give on, and eventually through all the give-and-take, a solution is reached. Neither side wins or loses in this situation. As a result, neither side really buys in to the decision that was reached. If, however, both parties make firm commitments to the resolution, then the solution can become a permanent one.

Confronting   *Confronting* is also called *problem-solving* and is the best way to resolve conflict. One of the key actions you'll perform with this technique is to go on a fact-finding mission. The thinking here is that one right solution to a problem exists and the facts will bear out that solution. Once the facts are uncovered, they're presented to the parties, and the decision will be clear. Thus, the solution becomes a permanent one, and the conflict expires. This is the conflict-resolution approach project managers use most often and is an example of a win-win technique.

Avoiding   *Avoiding*, sometimes known as withdrawal, never results in resolution. This occurs when one of the parties gets up and leaves and refuses to discuss the conflict. It is probably the worst of all the techniques because nothing gets resolved. This is an example of a lose-lose conflict-resolution technique.

Negotiating   *Negotiating* is a technique I've discussed before. This involves both parties communicating, listening, and asking questions. Sometimes negotiating uses a third party who has no vested interest in the outcome and is neutral regarding the solutions. This third party helps all parties reach an agreement. For example, negotiating is a technique often used by collective-bargaining organizations. Negotiating can be a win-win, win-lose, or lose-lose technique depending on how the negotiations are conducted and on the outcome.

These conflict management styles can help you understand the behaviors you're observing and help you reach resolution. Two additional situations that require special treatment are dealing with team member disputes and handling disgruntled team members, which I'll discuss next.

Team Member Disputes

Given the diverse backgrounds and varying areas of expertise among project team members, it should come as no surprise that team members will have disagreements. Sometimes people simply need to have a conversation and work through the issues, but other times disputes require the intervention of the project manager.

You may be tempted to make a snap judgment based on what you see at any given point in time, but this may only exacerbate the situation. You need to get the facts and understand what is behind the dispute. Interview each of the team members involved to get as much information as you can. If it's a minor dispute, you might consider hosting a meeting, with you as the moderator, and ask each person to explain their issues and offer potential solutions. You could turn this into a brainstorming session in order to engage everyone and place the burden of finding a solution on them.

Sometimes, the dispute is very deep or potentially involves threats or other workplace issues. Always get your human resource department involved in these issues as soon as you are made aware of the problem. Most organizations have strict policies and guidelines in place regarding disputes of this nature. They may recommend mediation, training, disciplinary action, or replacing one of the team members. Don't attempt to resolve these types of issues on your own. If you do, you may find yourself entangled in legal issues, especially if you acted outside of the company policy.

 It's always a good idea to check with your human resource department before jumping into the middle of dispute resolution. You want to make certain you are following company policies and don't end up as part of the problem yourself, rather than as part of the solution.

Disgruntled Team Members

Few situations can poison team morale more quickly than a disgruntled team member. This can happen at any time during the project and can involve anyone on the team.

The behavior of a discontented team member can take a variety of forms. They may become argumentative in meetings or continually make snide comments putting down the project. Even worse, this unhappy person may spend time "cube hopping" in order to share these negative feelings about the project with other team members. When otherwise-satisfied team members constantly hear statements that the project is worthless, is doomed to fail, or is on the cutting block, overall team productivity will be impacted.

As the project manager, you need to spend some private time with this employee to determine the cause of the dissatisfaction. It may be that the unhappy team member doesn't fully understand the project scope or how their contribution will lead to the project success. Or, it could be this person never wanted this assignment in the first place and feels forced onto a project they don't believe in.

It is best to start by listening. Stick to the facts, and ask the person to clarify the negative comments. If the team member is repeating incorrect information, set the record straight. If they are frustrated about some aspect of the project and feel no one is listening, find out what the issue is and explain that going around bad-mouthing the project is not the way issues get resolved. If the person truly does not want to be part of the project team or does not want to do their assigned tasks, work quickly with the functional manager or your sponsor to get this person replaced.

It's your responsibility as the project manager to hold your team members accountable. Once you've given them an opportunity to state their case and have made a few positive changes to address their concerns, make it clear you expect their negative behavior to stop. If they don't, you'll need to get your human resource experts involved and begin disciplinary action.

Ideally, conflict resolution should not dominate your time with the team. In my experience, this is typically a one- or two-time issue on most projects, depending on the length and complexity of the project. Most likely, you'll be more involved with building and managing a cohesive team and using an effective rewards and recognition system to motivate the team. You'll look at rewards and recognition systems next.

Now that you have the team established and assembled, it's time to have the project kickoff meeting.

Project Kickoff

A project kickoff meeting is generally held after the project charter is signed. However, in some cases you may not have all the project team members assigned at that time. If that's the case, it's a good idea to hold off on setting up the kickoff meeting until the majority of the team members are assigned.

It is also common to conduct two kickoff meetings to avoid this problem. One meeting is held after signing the project charter, and another one is held to kick off the Executing phase of the project once all team members are assigned.

> The timing of the kickoff meeting isn't as important as actually holding a kickoff meeting. It should occur early on in the project, ideally after the project charter is signed.

The project kickoff meeting is the best way to formally introduce team members and stakeholders and convey the same message to everyone at the same time. You may not know all of your team members, and you may not even have had the opportunity to interview them for the positions they will fill, depending on how they were selected (or appointed) to work on the project. As I've discussed, some organizations provide team members based on the functional manager's say-so, with little input from the project manager.

The tone that you set at the project kickoff meeting can make or break your relationship with the team. An ideal project kickoff session is a combination of serious business and fun. Your goal is to get the team aligned around the project goals and to get the team members comfortable with each other. This is a great opportunity to begin the forming stage.

There are many ways to structure a kickoff meeting. Here are some of the key components you may choose to include:

Welcome   It is a good idea to start the meeting by welcoming the team members and letting them know that you are looking forward to working with them. The welcome also gives you an opportunity to set the stage for the rest of the day. Take a few minutes to run through what participants can expect out of the meeting and what activities they will be involved in during the course of the project.

Introductions   A typical introduction format may include the person's functional area, brief background, and role in the project. The project manager should start the process to set an example of the appropriate length and detail. Put some thought into the information you want team members to share so that the time invested is worthwhile.

Project Sponsor and Key Stakeholders　Invite the project sponsor, the customer, and any other executive stakeholders who are key to the project. It's important that the team members know them and hear their goals and expectations for the project firsthand. These people may not be able to stay for the whole session, but do your best to get them to at least make an appearance and say a few words to the team.

You may need to do some coaching here, so spend time prior to the session communicating with the executive stakeholders regarding the message they should deliver. If your sponsor happens to be a dynamic speaker, you might want to schedule them for a little more time to get the troops excited about the project they are working on.

Project Overview　You'll start out this section with the project goals and objectives. You should also summarize the key deliverables for each of the project phases, as well as the high-level schedule and budget. This overview will help team members get the big picture and understand how they fit on the project. It also helps set the foundation regarding the purpose and goals for the project.

Stakeholder Expectations　This section is a natural segue from the previous section. Along with explaining the goals, schedule, and budget, it's important that the team understands the stakeholder expectations for the project. Explain the reasons for the project deadline or budget constraints if they exist. Make certain team members are aware of any quality concerns, political issues, or market announcements that are tied to this project.

Roles and Responsibilities　Start this section with a description of your roles and responsibilities for the project. Many of the team members may not know you or be familiar with your management style, so this is your chance to communicate how you will be managing the project and your expectations for how the team will function.

Depending on the size of the project, you may want to review the roles and responsibilities for each key team member or skill area. Let them know your expectations regarding project management procedures, reporting and escalation of issues, team meeting schedules, what you expect in terms of individual progress reports, and how they will be asked to provide input into project progress reports.

Question and Answer　One of the most important agenda items for the kickoff session is the time you allocate for team members to ask questions. This engages them on the project and is the ideal opportunity to clarify questions regarding goals, deliverables, expectations, and more.

 Real World Scenario

Kickoff for Remote Team Members

For a project kickoff to work effectively, it needs to include all team members. But what do you do if part of your team is located in a different city or state?

Remote team members often feel left out, especially if the majority of the team, including the project manager, sponsor, and client, is located at corporate headquarters where all of the action is.

Getting approval to bring in remote team members is a battle worth fighting, because it's so important in making everyone feel like part of the team. When making your case with the project sponsor, make sure you explain the importance of this meeting and the benefits it will have to the project. Your sponsor will be much more receptive to the idea if they know what will be covered and can see that this exercise is far more than people getting together for a free lunch.

But with more companies tightening their belts and looking closely at travel-related expenses, bringing in remote team members may not be possible. Perhaps if you can't get the budget to bring in all the team members, maybe you can do the next best thing: fly the sponsor, the project manager, and a key stakeholder or two to the place where the other team members reside and hold a separate kickoff for them. Last but not least, there are teleconferencing, video conferencing, and web-based meeting options that you can use to at least virtually bring everyone together during the kickoff meeting.

The kickoff meeting is an excellent opportunity to get everyone on the same page. At the time you start project execution, you will probably have a combination of people who have been involved with the project since initiation and those who are relatively new to the project. This meeting is the time to set expectations and remind everyone of their role on the project.

 **Real World Scenario**

Main Street Office Move: Building the Team

You've worked with the functional managers to select team members for your project. You created a project organization chart and also developed a RACI matrix. A sample portion of the RACI is shown here. Juliette heads up the communication department. Leah is the manager of the procurement department, Jason is the manager of information technology, Joe is the fleet manager, and Kate is the executive sponsor.

Deliverable/stakeholder	Juliette: Communication	Leah: Procurement	Jason: IT	Joe: Fleet	Kate: Sponsor
Communication	R	C	C	C	A/C
Moving company	I	A/R	C	I	C/I
Seating charts	A/R	I	C		C/I
Technical installs	I	I	A/R		C/I
Fleet cars	I	C	I	A/R	I

Since all the team members are new to the project and some have not worked together before, you decide to hold some team-building activities. Your human resource department assists you with some exercises that help to break the ice and allow team members to get to know each other. They progress through the forming stage of development quickly and stall a bit during the storming stage. This isn't unusual, and the team continues becoming more cohesive as the project continues.

A conflict arises between two of the team members concerning the choices of furniture and fixtures for the new office space. As soon as you realize the tension is building between these team members, you speak with each of them and use the confronting (problem-solving) conflict-resolution technique to reach agreement between them.

Summary

All projects require resources. Resources include the human type and physical resources such as material and equipment.

Shared resources work for more than one manager. They may report to both a functional manager and a project manager. If you're working with a shared resource, be certain you have some input into the team member's performance appraisal. Low-quality resources may not have the skills needed to complete the tasks assigned. In the case of low-quality physical resources, look elsewhere.

Resource allocation identifies the type of resources needed, skills sets, and time frames the skills are needed. Resource shortages occur when there aren't enough resources available with the skills needed to perform the task. Resource overallocation occurs when one resource (or set of resources) is scheduled to work on too many tasks at the same time. These tasks may be on the same project or a combination of tasks from different projects and their normal operational work. Benched resources are those who have rolled off one project and are waiting for the next project to begin. These are costly resources to the organization, so work in coordination with other project and functional managers to assign these resources as soon as possible.

Interproject dependencies occur when you need the completed deliverables from one project in order to start work on the next project. Interproject resources contention occurs when resources are scheduled for similar tasks on competing projects and may be unavailable for your project when needed.

Team members might be in-house resources, resources from outside the organization, or *remote resources*. When possible, colocate resources.

A project organization chart is a hierarchical chart that shows the sponsor, project manager, and team members. A RACI chart shows the roles and responsibilities of team members (or it can depict teams or business units). RACI stands for responsible, accountable, consulted, and informed.

All teams progress through five stages of development: forming, storming, norming, performing, and adjourning.

Team-building activities help diverse groups of people work together in an efficient and effective manner. Trust building is an important activity. It takes time to build trust with team members and requires that the project manager demonstrate competence, respect, honesty, integrity, and openness.

Provide feedback to your team members in a timely manner and deal with negative situations as soon as they occur. Removing personnel from the project should be done in coordination with the human resource department.

Rewards and recognition are important motivators for individuals and teams. Be certain the reward is in keeping with the achievement. Also make certain that reward criteria and processes are written down.

Conflict will occur on nearly all projects. The common techniques for conflict resolution include smoothing, forcing, compromising, confronting, avoiding, and negotiating. Confronting, also known as problem-solving, is the technique project managers should use.

Project kickoff meetings introduce the team members and stakeholders to each other and describe the goals and objectives of the project. They are typically held after the project charter is signed.

Exam Essentials

Understand the definition of resources.   Resources can be human resources or physical resources. They are used to complete the work of the project. Resources include these categories: shared resources, dedicated resources, low-quality resources, in-house resources, benched resources, and remote resources.

Define resource allocation.   Resource allocation is identifying resource availability and skill sets and assigning them to project tasks.

Define resource overallocation and resource shortage.   Resource overallocation occurs when resources are assigned too many tasks within a given time frame. Resource shortage occurs when there are not enough resources with the required skills or abilities to complete the tasks.

Define interproject dependencies and interproject resource contention.   Interproject dependencies occur when one project must complete its deliverables before another project can begin. Interproject resource contention occurs when resources are assigned to more than one project resulting in timing and availability conflict.

Define a RACI chart and define the acronym.   This is a matrix-based chart that shows the resource role and responsibility level for the work product. RACI stands for responsible, accountable, consulted, and informed.

Name the five stages of team development.   They are forming, storming, norming, performing, and adjourning.

Describe team building and trust building.   Team building consists of activities that help diverse groups of people work together in an efficient and effective manner. Trust building involves building trust with the project manager and among team members. This takes time and is accomplished by being true to your word and having the team's best interests at heart.

Name the conflict-resolution techniques and the technique that is best for project managers.   They are smoothing, forcing, compromising, confronting, avoiding, and negotiating. Confronting is also known as problem-solving and is the technique project managers should use.

State the purpose of a project kickoff meeting.   The project kickoff meeting is a way to formally introduce all project team members, to review the goals and the deliverables for the project, to discuss roles and responsibilities, and to review stakeholder expectations.

Key Terms

Before you take the exam, be certain you are familiar with the following terms:

adjourning

avoiding

benched resources

collocate

colocate

compromising

conflict

confronting

dedicated resources

forcing

forming

in-house resources

interproject dependencies

interproject resource contention

low-quality resources

negotiating

norming

organization breakdown structure (OBS)

performing

physical resource

problem-solving

project organization chart

RACI chart

remote resources

responsibility assignment matrix (RAM)

resource allocation

resource breakdown structure (RBS)

resource shortage

resource overallocation

shared resource

smoothing

storming

team building

trust building

Review Questions

1. Your project is underway and your team members are working well together, anticipating the needs of the project, and they all understand their roles in the project. A new team member has been introduced and started working with the team this week. Which stage of team development does this situation represent?

   A. Forming

   B. Storming

   C. Performing

   D. Norming

2. Which of the following helps to build an efficient and effective team, improves morale, and builds social bonds?

   A. Team building

   B. Trust building

   C. Having dedicated resources on the project team

   D. Using appropriate conflict-resolution techniques at the right time

3. A team member has come to your office to complain that a fellow team member is never available for meetings before noon and seems to ignore her requests to follow proper processes. Which of the following does this describe? Choose two.

   A. This describes an interproject resource contention.

   B. This describes a low-quality resource.

   C. This describes a conflict.

   D. This describes a situation where the negotiating conflict technique should be used.

   E. This describes varying work styles.

4. All of the following are stages of team development except for which one?

   A. Adjourning

   B. Confronting

   C. Performing

   D. Storming

5. One of the key project team members is assigned to two other projects. The project schedule shows this resource needed during the same time they're needed on one of the other projects. What does this describe?

   A. Resource allocation

   B. Interproject resource contention

   C. Resource shortage

   D. Resource conflict

6. Your project is undergoing some difficulties. You've determined that the primary problem is vendor performance. A key stakeholder insists the problem is not the vendor; the problem is the project team. However, the key stakeholder spends most of the meeting emphasizing the areas of agreement. Which conflict-resolution technique does this describe?

    A. Avoiding

    B. Forcing

    C. Confronting

    D. Smoothing

7. You have two stakeholders who are at odds about the scope of the project work. You listen, ask questions, and lead them both to agree that some scope items need to be added and others changed. Which conflict-resolution technique does this describe?

    A. Confronting

    B. Smoothing

    C. Negotiating

    D. Avoiding

8. Which of the following are conflict-resolution techniques? Choose three.

    A. Adjourning

    B. Negotiating

    C. Norming

    D. Avoiding

    E. Compromising

9. Your system engineer has started making negative comments during your weekly team meeting. He has had a heated argument with the marketing manager, and you have heard from various team members that he has become difficult to work with. What is the best course of action for you to take?

    A. You should write a memo to the system engineer's functional manager and request a replacement as soon as possible.

    B. The system engineer is critical to the project, so you should give him some slack and wait to see whether the behavior stops.

    C. You should confront the system engineer openly at the next team meeting. Let him know that his behavior is unacceptable and that he will be replaced if there is not an immediate change.

    D. You should schedule an individual meeting with the system engineer to determine whether he has issues with the project that need to be resolved. Get his perspective on how the project is progressing and how he feels about his role.

10. Which of the following is true regarding a RACI chart? Choose three.

    A. A RACI chart shows roles and responsibilities of team members or business units and how they intersect with project tasks.

    B. RACI stands for responsible, approved, consulted, and informed.

C. A RACI is a matrix-based chart.

D. A RACI is a type of organization breakdown structure.

E. RACI stands for responsible, accountable, consulted, and informed.

11. All of the following are true regarding rewards and recognition except for which one?

A. You should have a written procedure describing the criteria for rewarding team members.

B. Rewards and recognition are a form of motivation.

C. Rewards and recognition should be applied consistently to all project team members.

D. Rewards and recognition almost always involve money.

12. You are working on a construction project. Your organization owns one crane, and the crane is needed for two tasks on your project at the same time. This is known as which of the following?

A. Resource shortage

B. Resource allocation

C. Interproject dependencies

D. Shared resource

13. This meeting, held after the project charter is signed or at the beginning of the Executing phase, formally introduces the team members and stakeholders and outlines the goals and objectives of the project for the team.

A. Project status meeting

B. Kickoff meeting

C. Project introductory meeting

D. Steering-committee meeting

14. The project sponsor will approve the final deliverables of the project. On a RACI chart, how would this be designated?

A. R

B. I

C. A

D. C

15. Your team is working well together. They understand their roles on the project, perform the work with the best effort possible, and really enjoy working with one another. A new team member will be introduced next week who is well respected by the current team. Which of the following is true regarding this situation? Choose two.

A. The team is currently in the performing stage of team development.

B. Once the new team member is introduced, the team will revert to the forming stage of team development.

C. The team is currently in the performing stage of team development and will stay in performing because they know the new team member.

    D. Once the new team member is introduced, the team will revert to the norming stage of team development.

    E. The team is currently in the performing stage of team development and will revert to the storming stage of team development because they know the new team member.

16. Of the situations listed, for which would team-building efforts have the greatest impact? Choose three.

    A. Schedule changes

    B. Organizational changes

    C. Personality clashes

    D. Budget changes

    E. Staff changes

    F. Organizational changes

17. All the following are elements of the project kickoff meeting except for which one?

    A. Introductions

    B. Overview of goals and objectives

    C. High-level budget overview

    D. Roles and responsibilities overview

    E. WBS overview

    F. Stakeholder expectations

18. Which conflict-resolution technique is known as win-lose?

    A. Smoothing

    B. Avoiding

    C. Confronting

    D. Forcing

19. Resources who are awaiting new assignments between projects are costly to an organization and typically reside in a projectized organizational structure. Which of the following does this describe?

    A. Overallocated resources

    B. Benched resources

    C. In-house resources

    D. Remote resources

20. One way you are presenting the work of the project is by listing the departments responsible for the work along with the work packages they're assigned to. What type of chart is this?

    A. RBS

    B. Project organization chart

    C. OBS

    D. Organization chart

Chapter

7

Defining the Project Budget and Risk Plans

THE COMPTIA PROJECT+ EXAM TOPICS COVERED IN THIS CHAPTER INCLUDE

✓ **1.4   Identify the basics of project cost control.**

- Total project cost
- Expenditure tracking
- Expenditure reporting
- Burn rate
- Cost baseline/budget
  - Plan vs. actual

✓ **2.2   Explain the importance of risk strategies and activities.**

- Strategies
  - Accept
  - Mitigate
  - Transfer
  - Avoid
  - Exploit
- Risk activities
  - Identification
  - Quantification
  - Planning
  - Review
  - Response
  - Register
  - Prioritization
  - Communication

✓ **4.1   Compare and contrast various project management tools.**

- SWOT analysis

This chapter will cover cost estimating and cost budgeting, as well as the risk activities and strategies for your project.

To many project managers, the most important components of a project plan are the scope statement, schedule, and budget. As you probably recall, these also happen to constitute the classic definition of the triple constraints. This doesn't mean that other components of the plan aren't important. The communication plan and the risk management plan, for example, are also important on any project.

Estimating Costs

Now that you've developed the project schedule and documented the project resource requirements, you're ready to begin estimating the cost to complete the work of the project, determine the project budget, and estimate the total cost of the project.

Stakeholders have a way of declaring both cost and schedule estimates as final. Make certain they are aware that these are estimates and that more information will be known about total cost as the project progresses. Estimates become final once you determine the cost baseline.

You can use a number of techniques to determine cost estimates. I'll cover four in this chapter, including analogous (also known as *top-down*), parametric, bottom-up, and three-point estimates. I'll also provide some tips to help you work through the estimating process. Remember that these estimating techniques can be used for estimating activity durations as well as costs.

Cost-Estimating Techniques

You can use several techniques to estimate project costs; you may use some or all of these methods at various stages of project planning, or you may use one type of estimating for certain types of activities and another method for the rest.

The estimating methods have varying degrees of accuracy, and each method can produce different results, so it is important to communicate which method you are using when you provide cost estimates. Let's look at each of these estimating methods in more detail and discuss how they work.

Analogous Estimating (Top-Down Estimating)

I talked about analogous estimating when discussing schedule planning in Chapter 5. For cost-estimating purposes, an analogous estimate approximates the cost of the project at a high level by using a similar past project as a basis for the estimate. (You may also hear this technique referred to as *top-down estimating* or as making an *order-of-magnitude* estimate.) This type of estimate is typically done as part of the business case development or during the early stages of scope planning when there isn't a lot of detail on the project. Analogous estimating uses historical data from past projects along with expert judgment to create a big-picture estimate. Remember that expert judgment relies on people who have experience working on projects of similar size and complexity or who have expertise in a certain area. An analogous estimate may be done for the project as a whole or for selected phases or deliverables. It is not typically used to estimate individual work packages.

Analogous estimates are the least accurate of all the estimating techniques but also the least costly. Analogous estimates rely on expert judgment.

Analogous estimating will likely be the best technique to use at the early stage of the project because you'll have very little detail to go on. The key here is to make sure that everyone involved understands how imprecise this estimate is.

Parametric Estimating

Chapter 5 discussed parametric estimating. You'll recall that this technique uses a mathematical model to compute costs, and it most often uses the quantity of work multiplied by the rate. Also, commercial parametric modeling packages are available for complex projects or those performed within specialized industries.

Parametric estimating is dependent on the accuracy of the data used to create the model. If your organization uses parametric modeling, spend some time learning about the specific models that are available and whether this technique is appropriate for your specific project.

Bottom-Up Estimating

The most precise cost-estimating technique is called the *bottom-up estimate*, which assigns a cost estimate to each work package on the project. The WBS and the project resource requirements are critical inputs for a bottom-up estimate. The idea is that you start at the work package level of the WBS and calculate the cost of each activity assigned to that work package. The sum of all the work package estimates provides the estimate of the total project cost.

Bottom-up estimates are the most accurate of all the estimating techniques, but they're also the most time-consuming to perform.

When I discussed schedule planning in Chapter 5, I talked about calculating duration estimates for each task to determine the length of time your project will take. When you are calculating cost estimates, you need to base the estimate on *work effort*, which is the total time it will take for a person to complete the task if they do nothing else from the time they start until the task is complete. For example, assume a task to perform technical writing has an activity duration estimate of four days. When you perform cost planning, you need to know the actual number of hours spent performing the task. So, let's say the technical writer is allocating 5 hours a day to the project over the course of 4 days. The work effort estimate is 5 hours a day multiplied by 4 days, which equals 20 hours.

The duration estimates that you complete in schedule planning help you define how long the project will take to complete. The work effort estimates that you obtain in cost planning are used to define how much the project will cost.

The final piece of data you need for a bottom-up estimate is the rate for each resource. Rates for labor and leased equipment are typically calculated on an hourly or daily rate, while the purchase of materials or equipment will generally have a fixed price.

Deciding the correct rate to use for cost estimates can be tricky. For materials or equipment, the current cost of a similar item is probably as accurate as you can get. The largest cost for many projects is the human resource or labor cost. The actual rate that someone will be paid to perform work, even within the same job title, can fluctuate based on education and experience level. If you're procuring contract resources, you'll get a rate sheet that describes the rates for a given job title and the resource's travel rates if they're coming from out of town.

Table 7.1 shows the work effort and rate assigned for each of the resources in a sample project.

TABLE 7.1   Sample project resource rates

Task	Resource	Work Effort	Rate
4.1.1	Tech writer	20 hours	$30/hr
4.1.2	Programmer	100 hours	$150/hr
4.1.3	Server	Fixed rate	$100,000
4.1.4	Testers	60 hours	$80/hr
4.1.5	Programmer	200 hours	$150/hr
4.1.6	Marketing	30 hours	$60/hr

Now that you have the resource requirements, the work effort estimates, and the rate for each task, you can complete the cost estimate by adding a Total Cost column to the table. The cost of each task is calculated by multiplying the work effort for each resource by the rate for that resource. This will give you the total project cost estimate. Table 7.2 shows a completed cost estimate for the tasks in the sample project.

TABLE 7.2 Sample project cost estimate

Task	Resource	Work Effort	Rate	Total Cost
4.1.1	Tech writer	20 hours	$30/hr	$600
4.1.2	Programmers	100 hours	$150/hr	$15,000
4.1.3	Server	Fixed rate	$100,000	$100,000
4.1.4	Testers	60 hours	$80/hr	$4,800
4.1.5	Programmer	200 hours	$150/hr	$30,000
4.1.6	Marketing	30 hours	$60/hr	$1,800
Total				**$152,200**

Three-Point Estimates

Three-point estimates are an average of the most likely estimate, the optimistic estimate, and the pessimistic estimate. You should ask team members and/or ask subject-matter experts who are familiar with this type of purchase or consulting work to give you estimates. They should base these on their past experience with similar work or their best guess based on their expertise.

The most likely estimate assumes that costs will come in as expected. If you are using this to estimate activity durations, this estimate assumes work proceeds according to plan, that there won't be any obstacles, and that the team member is confident they have the skills to complete the task.

The optimistic estimate is an estimate that is better than expected. For example, you may have estimated internally that the consulting work you are requesting will be $225 per hour. The optimistic estimate might be $212 per hour, which is better than what you expected. If you are using this estimate for activity duration, this estimate is the fastest time frame in which your resources can complete the activity. This might assume that other tasks the resource is working on are completed early or no longer need to be worked on during the same time frame as the current task. It could also mean that the work is easier than anticipated so the task could complete early.

The pessimistic estimate assumes the goods or services will cost more than expected. If you are using this for activity duration, it assumes the work will take longer than anticipated to complete or that obstacles will crop up along the way that will delay completing the work.

Calculating the three-point estimate is straightforward. It's an average of the sum of the estimates. For example, let's say your most likely estimate is $120 per hour. The optimistic estimate is $110 per hour, and the pessimistic estimate is $150 per hour. The three-point estimate is calculated this way:

($120 + $110 + $150) / 3 = $126.67 per hour

The three-point estimate for this work is $126.67 per hour. This estimate is only as good as the estimates given for the three points in the calculation. This estimate can swing between accurate and poor based on the quality of the estimates provided by the subject-matter experts.

> Most of the estimating techniques discussed in this section can also be used to determine the task durations on the project schedule.

Estimating Tips

Cost estimating can be complex, and cost estimates often turn into the official cost of the project before you have the proper level of detail. You will probably never have all the information that you'd like when calculating cost estimates, but that's the nature of project management. Here are some thoughts to keep in mind as you work through the estimating process:

Brainstorm with your project team. Work with your team and other subject-matter experts to make certain you've accounted for cost estimates that may not be so obvious. For example, do any of your project team members require special training in order to perform their duties on the project? Are there travel costs involved for team members or consultants? Getting the team together to talk about other possible costs is a good way to catch these items.

Communicate the type of estimate you are providing. Make certain your stakeholders are aware of the types of estimates you're using and the level of accuracy they provide. If you're preparing an analogous estimate based on a similar project, be up front regarding the possibility of this estimate deviating from the actual cost of the project.

> In addition to emphasizing the potential inaccuracies of an analogous estimate, provide stakeholders with a timeline for a definitive estimate. A project sponsor is more willing to accept that your current estimate may be lower than the actual cost of the project if they understand why the current estimate is vague and what is being done to provide a more accurate estimate.

Make use of any available templates. Many companies have cost-estimating templates or worksheets that you can use to make this job easier. They will also help you in identifying hidden costs you may not have thought of and may list the rates from the vendor agreements the organization has entered into.

Templates may also be a good source of rate estimates for internal resources. The salary of the people on your project will vary based on both their job title and specific experience. Your organization may require you to use a loaded rate, which is typically a percentage of the employee's salary to cover benefits such as medical, disability, or pension plans. Individual corporate policies will determine whether loaded rates should be used for project cost estimates.

Get estimates from the people doing the work. A bottom-up estimate is the most accurate because effort estimates are provided for each activity and then rolled up into an overall estimate for the deliverable or the project. The person performing the activities should be the one to develop the estimate. If your project includes tasks new to your team or uses an untested methodology, you may need to look outside for assistance with work effort estimates. You could consult published industry standards or hire a consultant to assist with the estimating process.

Document any assumptions you have made. Make certain to document any assumptions you've made when performing cost estimates. For example, you may need to note that you are assuming the rate sheet you're using to determine contractor costs will still be valid once the work of the project begins.

The cost estimates will be used to create the project budget, which you'll learn about next.

Creating the Project Budget

When you have the cost estimates completed, it's time to prepare the budget. *Budgeting* is the process of aggregating all the cost estimates and establishing a cost baseline for the project. The *cost baseline* is the total expected cost for the project. Once approved, it's used throughout the remainder of the project to measure the overall cost performance.

The project budget is used to track the actual expenses incurred against the estimates. You'll look at tracking and reporting expenses more closely in the "Expenditure Tracking and Reporting" section later in this chapter.

Before learning about the mechanics of the budget itself, you should make sure you have an understanding of the processes within your organization regarding budgets, authority levels, how expenses are approved, and more. Here are a few questions you can use to help get you started:

- Are all project expenses submitted to the project manager for approval?
- What spending authority or approval levels does the project manager have regarding project expenses?

- Does the project manager approve time sheets for project team members?
- Are there categories of cost or threshold dollar amounts that require approval from the project sponsor or customer?

Getting the answers to these questions before spending the money will eliminate problems and confusion later in the project.

Tracking project expenses as they're incurred is not always the responsibility of the project manager. Once the cost estimates have been provided and the project budget is established, the actual tracking of expenses may be performed by the accounting or finance department. Some organizations use their program management office (PMO) to oversee project budgets, approve expenses, track all the project budgets, and so on. Make certain you know who is responsible for what actions regarding the budget.

No matter who actually tracks the budget expenditures, as the project manager, you're the person accountable for how the money is spent and for completing the work of the project within budget. You'll want to monitor the budget reports regularly so you can identify any overruns and take corrective action to get the budget back on track.

> Set up a routine meeting with all the budget analysts from the various departments providing project funding so that everyone is aware of the status of the project budget at any point in the project.

Now let's look at how to create the project budget.

Creating the Project Budget

Project budgets are usually broken down by specific cost categories that are defined by the accounting department. A few examples of common cost categories include salary, hardware, software, travel, training, and materials and supplies. Make certain to obtain a copy of your organization's cost categories so that you understand how each of your resources should be tracked and classified.

> Check with your PMO or accounting department to see whether there are standard budget templates that you can use for the project.

Project budgets are as varied as projects themselves. Although the format for budgets may be similar from project to project, the expenses, budget amounts, and categories you use will change for each project.

Most budgets are typically created in a spreadsheet format and may be divided into monthly or quarterly increments or more depending on the size and length of the project. If you don't have a template available to start your budget, contact your accounting department to get the chart of accounts information needed to construct the budget. The chart of accounts lists the account number and description for each category of expense you'll use on the budget. You may remember that we used the chart of accounts codes earlier in the book for the work package level of the WBS as well.

To begin creating the budget, list categories such as salaries, contract expenses, materials, travel, training, and others, and record the cost estimates you derived for each. Add a column for actual expenditures to date. You'll use this information during the Monitoring and Controlling phase of the project to determine the financial health of the project. Table 7.3 shows a high-level sample budget.

TABLE 7.3 Sample project budget

Account Code	Category	Estimated Costs
1001	Contract labor	$50,000
1003	Materials	$2,500
1005	Hardware	$22,700
1010	Training	$7,000
Total Budget		**$82,200**

You may want to include two additional types of expenses in the project budget: contingency reserves and management reserves. Make certain you check with your organization regarding the policies that dictate the allocation of these funds and the approvals needed to spend them.

Contingency Reserve A *contingency reserve* is a certain amount of money set aside to cover costs resulting from possible adverse events or unexpected issues on the project. These costs may come about for many reasons, including scope creep, risks, change requests, variances in estimates, cost overruns, and so on. There is no set rule for defining the amount you should put in a contingency fund, but most organizations that use this allocation often set the contingency fund amount as a percentage of the total project cost.

Be aware that stakeholders may misunderstand the meaning of a contingency reserve and see it as a source of funding for project enhancements or additional functionality they didn't plan into the project. Make certain they understand the purpose of this fund is to cover possible adverse or unexpected events. With the exception of very small projects, it seems there are always expenses that come up later during the project that weren't planned for up front. The contingency fund is designed to cover these types of expenses.

Management Reserve A *management reserve* is an amount set aside by upper management to cover future situations that can't be predicted. As with the contingency reserve, the amount of a management reserve is typically based on a percentage of the total project cost.

What makes the management reserve different from the contingency reserve is the spending authority and the fact that it covers unforeseen costs. The contingency fund is usually under the discretion of the project manager, who controls how these funds are spent. The management fund is usually controlled by upper management, and the project manager can't spend money out of this fund without approval from upper management. Management reserves are not included as part of the project budget or cost baseline.

The terms *contingency reserve* and *management reserve* may be considered interchangeable in some organizations.

The project budget is used to create the cost baseline, which is a tool used during project execution and during the Monitoring and Controlling phase to monitor project expenditures.

Cost Baseline

The key members of the project team should review the draft budget. It may be appropriate to have a representative from the accounting department or the PMO review the draft as well. The project team needs to understand the critical link between the schedule and the budget. Any questions about budget categories or how the dollars are spread across the project timeline should be addressed at this time.

Once the budget review with the project team is complete, it is time to get the project sponsor's approval and then create a cost baseline. The cost baseline is the total approved expected cost for the project. This should be approved before any work begins. All future expenditures and variances will be measured against this baseline.

The cost baseline is used throughout the remainder of the project to track the actual cost of the project against the estimated or planned costs. It is also used to predict future costs based on what's been spent to date and to calculate the projected cost of the remaining work.

The project manager should communicate the cost baseline to the project stakeholders. Some stakeholders may want a copy of the total project budget, while others may be interested in the budget only for specific phases of the project. You should note each stakeholder's needs regarding budget information in the communication plan.

Expenditure Tracking and Reporting

Now that the budget is established and the cost baseline is approved, you will need to track the project expenditures and report the state of the budget to your stakeholders and sponsor. It's common to include budget updates as an agenda item during your project update meetings.

Expenditure tracking includes measuring the project spending to date, determining the *burn rate* (or how fast you're going through the money), and tracking expenditures to the cost baseline so that stakeholders can see what was planned versus what was actually spent on the project. *Expenditure reporting* is the mechanism you'll use to report on the current state of the project budget.

Project management software is useful in tracking project spending. You can run reports that show spending to date versus what was planned, and you can also use software to look at the impact of adding new tasks or resources using what-if scenarios.

You can also use your handy spreadsheet program for expenditure tracking and reporting. Table 7.4 is a sample of a budget report showing the estimated costs (what was planned) versus the actual costs to date and the variance, or difference between the two.

TABLE 7.4   Sample project budget report

Account Code	Category	Estimated Costs	Actual Cost at Reporting Date	Variance
1001	Contract labor	$50,000	$48,500	$1,500
1003	Materials	$2,500	$2,500	$0
1005	Hardware	$22,700	$24,500	$(1,800)
1010	Training	$7,000	$5,000	$2,000
Total Variance This Period		$82,200	$80,500	**$1,700**

Earned Value Management

One of your duties as a project manager is to control and monitor the costs of the project. There are several techniques you can use to do this. *Earned value measurement (EVM)* is a performance measurement technique that compares what your project has produced to what you've spent by monitoring the planned value, earned value, and actual costs expended to produce the work of the project. When variances that result in changes to the cost baseline are discovered, those changes should be managed using the project change control system. If the budget needs updating and/or funding has been added to the project, you'll need approval and sign-off from the sponsor for the new cost baseline.

The primary functions of EVM analysis techniques are to determine and document the cause of the variance, to determine the impact of the variance, and to determine whether a corrective action should be implemented as a result.

EVM looks at schedule, cost, and scope project measurements together and compares them to the actual work completed to date. EVM is usually performed on the work packages or other WBS components. To perform the EVM calculations, you need to first gather these three measurements: planned value, actual cost, and earned value.

Planned Value   The *planned value (PV)* is the cost of work that has been authorized and budgeted for a specific schedule activity or WBS component (such as a work package) during a given time period or phase.

Actual Cost   *Actual cost (AC)* is the actual cost of completing the work component in a given time period. Actual costs might include direct and indirect costs but must correspond to what was budgeted for the activity. For example, if the budgeted amount did not include indirect costs, do not include them here.

Earned Value   *Earned value (EV)* is the value of the work completed to date as it compares to the budgeted amount (PV) for that period. EV is typically expressed as a percentage of the work completed compared to the budget. For example, if our budgeted amount is $1,000 and you have completed 30 percent of the work so far, your EV is $300. EV cannot exceed the total project budget, but it may exceed the PV for a work component or period of performance if the team is ahead of schedule because they completed work planned for a future period during the measurement period.

The concepts of PV, AC, and EV are really easy to mix up. In their simplest forms, here's what each means:

PV   The approved budget assigned to the work to be completed during a given time period

AC   Money that's actually been expended during a given time period for completed work

EV   The value of the work completed to date compared to the budget

Cost Variance

Cost variance (CV) tells you whether your costs are higher than budgeted (with a resulting negative number) or lower than budgeted (with a resulting positive number). It measures the actual performance to date against what's been spent.

The formula for CV is as follows:

$$CV = EV - AC$$

Suppose that as of December 1 (the measurement date), the performance measurements are as follows:

PV = 75,000

AC = 71,000

EV = 70,000

Now let's calculate the CV using these numbers:

70,000 − 71,000 = −1,000

The result is a negative number, which means that costs were higher than what was planned for the work that was completed as of December 1. These costs are usually not recoverable. If the result was a positive number, it would mean you spent less than what you planned for the work that was completed as of December 1.

Schedule Variance

Schedule variance, another popular EVM variance, compares an activity's actual progress to date to the estimated progress and is represented in terms of cost. It tells you whether the schedule is ahead of or behind what was planned for this period. This formula is most helpful when you've used the critical path methodology to build the project schedule. The schedule variance (SV) is calculated as follows:

$$SV = EV - PV$$

Plug in the numbers:

$$70,000 - 75,000 = -5,000$$

The resulting schedule variance is negative, which means you are behind schedule or behind where you planned to be as of December 1.

Together, the CV and SV are known as efficiency indicators for the project and can be used to compare the performance of all the projects in a portfolio.

Performance Indexes

Cost and schedule performance indexes are primarily used to calculate performance efficiencies, and they're often used to help predict future project performance.

The *cost performance index (CPI)* measures the value of the work completed at the measurement date against the actual cost. It is one of the most important EVM measurements because it tells you the cost efficiency for the work completed to date or at the completion of the project. If CPI is greater than 1, you are spending less than anticipated at the measurement date. If CPI is less than 1, you are spending more than anticipated for the work completed at the measurement date.

The cost performance index (CPI) is calculated this way:

$$CPI = EV \,/\, AC$$

Plug in the numbers to see where you stand:

$$70,000 \,/\, 71,000 = .99$$

Since the result is less than 1, it means cost performance is worse than expected.

The *schedule performance index (SPI)* measures the progress to date against the progress that was planned. This formula should be used in conjunction with an analysis of the critical path activities to determine whether the project will finish ahead of or behind schedule. If SPI is greater than 1, your performance is better than expected, and you're ahead of schedule. If SPI is less than 1, you're behind schedule at the measurement date.

The schedule performance index (SPI) is calculated this way:

$$SPI = EV \,/\, PV$$

Again, let's see where you stand with this example:

$$70,000 \,/\, 75,000 = .93$$

Schedule performance is not what you expected as of December 1.

Keep in mind that all of these formulas start with EV. Remember that the variances use subtraction and the performance indexes use division. Cost performance uses actual costs, and schedule performance uses planned value. Here's a recap of the formulas:

CV EV – AC

CPI EV/AC

SV EV – PV

SPI EV/PV

Burn Rate

Burn rate is the rate you are spending money over time. For example, if you had a $12,000 budget and you're spending $2,000 a month, you'll run out of money in six months. That's great if you are running a five- or six-month project. It's not so great if you're running a twelve-month project. Unfortunately, spending is rarely spread out evenly over the course of the project, so calculating burn rate isn't quite as simple as I've just explained. Burn rate on a project is typically calculated using the cost performance index (CPI) calculation.

Another method of determining burn rate uses the *estimate to complete (ETC)* formula. This is the cost estimate for the remaining project work. It is typically provided by the team members actually working on the project activities. Using the previous example, the project budget is $12,000, you've spent $8,000 to date, and the ETC is $6,000. Your spending is outpacing the project work, and you will likely run out of money unless corrections are made.

For the exam, remember that the cost processes include cost estimating, creating the project budget, and controlling costs.

Expenditure Reporting

Expenditure reporting can take many different forms. As shown earlier in Table 7.4, a simple spreadsheet showing planned versus actual costs to date and their variance should suffice for most small projects. Project management software gives you many options for tracking and reporting on the budget. Pie charts, bar charts, and budgets broken down by work components are other ways to report on and display project spending. You can also use much more advanced EVM measurements to report on project expenditures.

When project costs are displayed graphically over time, they form an S curve, as you can see in Figure 7.1. That's because at the beginning of the project, spending is minimal. It picks up once the work of the project is underway, and it tapers off at the end of the project as the work wraps up.

FIGURE 7.1   S curve

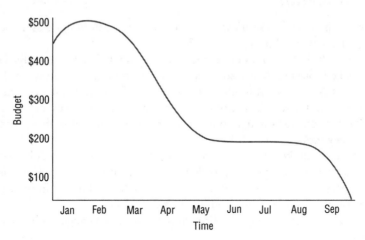

The important thing to know about budgeting is that you have a budget. Take the time to define estimates, create an overall budget based on the estimates, and then track what you're spending against what you planned and report this status to the stakeholders. Be certain to alert your sponsor at the earliest signs of trouble with the budget. You may be able to take action to get the budget back on track, but don't assume this can or will happen. The sponsor isn't likely going to be happy with negative budget news, but the sooner you make them aware, the sooner they can assist in solving the issue. Spending overruns are often outside the project manager's control, but you'll need documentation and budget reports when you talk with the sponsor.

Risk Planning

Risk is something that we deal with in our everyday lives. Some people seek out jobs or leisure activities that are considered high risk. They may get a thrill or a feeling of great accomplishment from taking on the challenge of skydiving or mountain climbing or from working as a lineman on high-voltage electrical lines.

A *risk* is a potential future event that can have either negative or positive impacts on the project. However, if you mention the word *risk* in association with a project, the majority of people will immediately think of something negative. In my experience, the vast majority of the time that a risk event occurs, it has negative consequences. But risks are not always negative. There is the potential for positive consequences as a result of a risk occurrence.

Risk planning deals with how you manage the areas of uncertainty in your project. There are three major components to risk planning: identifying the potential risks to your project, analyzing the potential impact of each risk, and developing an appropriate response for those risks with the greatest probability and impact. Let's take a look at each of these next.

Risk Identification

All projects have risks. *Risk identification* is the process of determining and documenting the potential risks that could occur on your project.

You can view risks both by looking at the project as a whole and by analyzing individual components of the project such as resources, schedule, costs, tasks, and so on. Risks can include such items as the level of funding committed to the project, the overall experience level of the core project team, the use of project management practices, or the strategic significance of the project. Risks may also be associated with particular phases of the project or with certain key tasks.

You should include stakeholders, core team members, and any other subject-matter experts who may have experience or knowledge of this project in your risk identification and analysis process. You can use several techniques to help define the initial risk list, including brainstorming, interviews, and facilitated workshops.

One way to get a jump start on a brainstorming session is to review a list of common potential risks:

- Budgets or funding
- Schedules
- Scope
- Requirements changes
- Contracts
- Hardware
- Political concerns
- Management risks
- Legal risks
- Technical issues

In addition, you could also devise a set of questions to pose to the group. Sample questions include the following:

- Is the task on the critical path?
- Is this a complex task?
- Does the task involve a new or unfamiliar technology?
- Does the task have multiple dependencies?
- Have you had problems with similar tasks in previous projects?
- Is this task controlled by outside influences (permits, regulatory compliance, environmental factors, and so on)?
- Are there inexperienced resources assigned to this task?
- Are there adequate resources assigned to the task?
- Are you unfamiliar with the hardware or software you're going to use for the tasks?

You can conduct the brainstorming meeting in a couple of ways. You can provide everyone with a marker and a sticky-note pad and ask them to write one risk per sticky note as they review the risk types and the questions. You can post the risks to a whiteboard and begin to categorize them. You can also invite someone to attend the meeting and take notes. No matter which method you use to identify the risks, at the conclusion of the meeting you should have a completed *risk register.*

This is essentially a list of risks that you can record in a simple spreadsheet that includes an identification number, risk name, risk description, risk owner, and response plan (or where the response plan can be located). The risk owner is the person responsible for monitoring the project to determine whether the potential for this risk event is high and for implementing the *risk response plan* should it occur.

Once you have identified all the possible risks to the project, you need to analyze the risks to determine the potential for the risk event to occur and the impact it will have on the project if it does occur.

What Could Possibly Go Wrong?

Another brainstorming technique you can use to identify risks is to hand out sticky notes and ask the risk team to answer one question: what could go wrong?

By letting people freely think and blurt out all the possibilities that occur to them, you may get some input that proves to be valuable in the risk identification process.

Strengths, Weaknesses, Opportunities, and Threats (SWOT)

Another technique you can use to identify risks is called *SWOT analysis.* This involves analyzing the project from each of these perspectives: strengths, weaknesses, opportunities, and threats. It also requires examining other elements such as the project management processes, resources, the organization, and so on.

Strengths and weaknesses are generally related to issues that are internal to the organization. Strengths examine what your organization does well and what your customers, or the stakeholders, view as your strengths. Weaknesses are areas the organization could improve upon. Typically, negative risks are associated with the organization's weaknesses and positive risks are associated with its strengths.

Opportunities and threats are usually external to the organization. For example, the weather, political climates in other countries where you may be performing parts of the project, the financial markets, and so on are external to the project and could present either opportunities or threats. SWOT analysis can be used in combination with brainstorming techniques to help discover and document potential risks.

Risk Analysis

Risk analysis is the process of identifying those risks that have the greatest possibility of occurring and the greatest impact to the project if they do occur. Before you begin the analysis process, it's important to understand your stakeholders' risk tolerance levels. This may be partially based on the type of industry you work in, corporate culture, departmental culture, or individual preferences of stakeholders. For example, research and development industries tend to have a high risk tolerance, while the banking industry leans toward a very low risk tolerance. Make certain you understand the risk tolerances of your stakeholders before assigning probability and impacts to the risk events.

You'll want to prioritize and quantify the risks in a way that is easy to understand and that highlights at a glance those risks that will need a risk response plan should they occur. One of the easiest ways to do this is by creating a *probability and impact matrix*. Now that the risk list is complete, the next step involves identifying the probability of the risk occurring and the impact.

Risk Probability   *Probability* is the likelihood that a risk event will occur. For example, if you a flip a coin, there's a .50 probability the coin will come up heads. Probability is expressed as a number from 0.0 to 1.0, with 1.0 being an absolute certainty the risk event will occur.

The quickest way to determine probability is by using expert judgment. This typically comes from knowledgeable project team members, stakeholders, or subject-matter experts who have experience on similar projects. Ask them to rate the probability of occurrence of each risk on the list. You could also review historical data from past projects that are similar to the current project to rate the probability of risk occurrence.

Risk Impact   *Impact* is the consequence (or opportunity) the risk poses to the project if it occurs. Some risks have impacts that are very low and won't impact the overall success of the project if they were to occur, while others could have impacts that cause a delay in the project completion or cause a significant budget overrun. Again, rely on the expert knowledge and judgment of the team members and on historical data to rate the severity of each risk. You can use a simple rating like the following to rate risk impact:

- High
- Medium
- Low

Next, you'll convert the high, medium, and low scores to a number between 0.0 and 1.0, with a score of 1.0 having the highest impact to the project, so that you can plug the scores into the probability and impact matrix. For example, a high impact may have a score of 1.0, a medium impact may have a score of 0.5, and a low impact may have a score of 0.1.

Check with your PMO to determine whether there are templates available that have predefined ratings for risk impacts.

Next you'll construct a probability and impact matrix to calculate a final risk score for each of the risks on your list. The final score is determined by multiplying the probability by the impact. Table 7.5 shows a basic probability and impact matrix.

TABLE 7.5   Probability and impact matrix

Risk	Probability	Impact	Risk Score
Risk A	0.9	1.0	0.9
Risk B	0.3	0.5	0.15
Risk C	0.8	0.1	0.08
Risk D	0.5	0.5	0.25

The closer the risk score is to 1.0, the more likely the risk will occur and have a significant impact on the project. In Table 7.5, risk A has a high probability of occurring and has a significant impact. Risk C has a high probability of occurring but a low impact if it does occur.

Make certain you understand the risk tolerances of your stakeholders before assigning probability and impacts to the risk events. If your stakeholders are primarily risk-averse but the team is assigning probability and impact scores that are allowing for a higher tolerance toward risks, when a risk event occurs, your stakeholders may react in a way you didn't expect. For example, if they perceive a certain risk as high and the team rated it as medium and then the risk event comes about, you may have stakeholders recommending canceling the project because of their unwillingness to deal with the risk event. Stakeholder reactions may bring about unintended consequences on the project, so be certain you're in tune with their comfort level regarding risk.

The last step in the risk process is to create an appropriate course of action for those risks with the highest scores.

Risk Response

Risk is, after all, uncertainty. The more you know about risks and their impacts beforehand, the better equipped you will be to handle a risk event when it occurs. The processes that involve risk concern balance. You want to find that point where you and the stakeholders are comfortable with the risk based on the benefits you can potentially gain. In a nutshell, you're balancing the action of taking a risk against avoiding the consequences or impacts of a risk event, or enjoying the benefits it may bring.

Risk response planning is the process of reviewing the risk analysis and determining what, if any, action should be taken to reduce negative impacts and take advantage of opportunities as a result of a risk event occurring.

Your organization may have a predetermined formula for identifying risks that require a response plan. For example, they may require that all risks with a total risk score greater than 0.6 must have a response plan.

You'll use several strategies when determining both negative risks and opportunities and formulating your response plans. The strategies to deal with negative risks include the following:

- *Avoid*: Avoiding the risk altogether or eliminating the cause of the risk event
- *Transfer*: Moving the liability for the risk to a third party by purchasing insurance, performance bonds, and so on
- *Mitigate*: Reducing the impact or the probability of the risk
- *Accept*: Choosing to accept the consequences of the risk

 Real World Scenario

You decide to take a road trip. However, you've had some rather unfortunate driving incidents lately. You've received two speeding tickets and were involved in a minor fender bender all within the last six months. You don't want to take a chance on another ticket, because your insurance rates are already astronomical. So, you hire a driver to go with you. The driver will do all the driving while you sit back and enjoy the scenery. You consider yourself very clever because you've transferred the risk of receiving a speeding ticket to the driver.

Your driver checks the road conditions and traffic reports before you leave. The driver discovers that the highway you planned to take is clogged with congestion because of an accident. You both decide to avoid the risk by taking an alternate route that will go around the accident and put you on the highway a few miles down the road, avoiding the congestion and the accident altogether.

A few days into the road trip, you find yourself on a mountain pass in the middle of a beautiful sunny day. All of a sudden, a rock tumbles down the side of the hill and lands squarely on the road in front of your car. The rock is so big you can't go around it. You and your driver accept this risk occurrence and perform a U-turn and go back to the town you just passed a few miles ago and wait for the road crew to clear the pass.

The strategies associated with positive risks or opportunities include the following:

- *Exploit*: Looking for opportunities to take advantage of positive impacts
- *Share*: Assigning the risk to a third party who is best able to bring about the opportunity
- *Enhance*: Monitoring the probability or impact of the risk event to assure benefits are realized
- Accept: Choosing to accept the consequences of the risk

As you perform risk analysis, assign owners, and determine whether response plans are needed, you should record this information in the *risk register*. Table 7.6 shows a sample risk register.

TABLE 7.6   Risk register

Risk #	Risk Description	Probability	Impact	Risk Score	Response Plan	Risk Owner	Response Plan
1	Bad weather	1.0	1.0	1.0	Y	Peterson	H Drive
2	Vendor delays	0.3	0.5	0.15	N	Hernandez	H Drive
3	Budget overrun	0.8	0.1	0.08	Y	Whatley	H Drive
4	Technical issues	0.5	0.5	0.25	N	White	IT Dept

The risk owner is responsible for monitoring the risks assigned to them and watching for risk triggers. *Risk triggers* are a sign or a precursor signaling that a risk event is about to occur. If the risk requires a response plan, the owner should be prepared to put the response plan into action once a trigger has surfaced.

Risk Monitoring

You will need to monitor risks throughout the Executing and Monitoring and Controlling phases of the project.

Risk events have a higher probability of occurring and more impact early on in the project, while probability and impact decrease as you get closer to closing out the project.

Use the risk register to communicate the project risks and action plans to the stakeholders. Include a risk update in your project status meetings, and periodically set up a risk review meeting to go through each risk on the risk register and determine whether they have occurred or are likely to occur. Risk typically diminishes over time, so risk rankings should change as well to reflect the current state of the project. New risks can come to light as you proceed with the work of the project, so be certain to make risk review a standard part of your project management process.

If you're working on a large project or one with an extended time frame, it's good practice to reevaluate risks periodically. This involves starting again at the risk identification process, working through the probability and impact exercise and writing response plans for newly identified risks, and updating responses for existing risks.

🌐 Real World Scenario

Main Street Office Move

Kate is anxiously awaiting the estimated cost for the project. Now that the schedule is complete and you have resources assigned, you begin the process of estimating the costs of the tasks, including contract resources, supplies, and materials. The work packages are a great place to start. Your estimates are here:

Description	Work Effort/ Quantity	Rate/Each	Estimated Cost	Comments
Communicate move purpose (marketing materials)	2,400 brochures	$1.50	$3,600	Estimate provided by Juliette in marketing
Moving company services	75 loads	$1,500	$112,500	Estimate based on calls and quotes from several moving companies
Furniture and fixtures	Desks, chairs, tables		$150,000	Estimate based on current pricing with the organization's vendor
Interior design	700 hours	$180/hr	$126,000	Estimate provided by facilities manager
Network printers	3	$25,000	$ 75,000	Estimate provided by Jason, IT director
Relocate fleet cars	16 hours	$75/hr	$1,200	Estimate provided by Joe, fleet manager
TOTAL Estimated Cost			**$468,300**	

Early in the project, Kate informed you that the budget was fixed at $450,000. The project estimates are coming in over that amount—not by much, but you know that this budget is a constraint and can't go over.

You also haven't accounted for contingency reserves in case there are unexpected costs. You will meet again with Leah in procurement to see whether she can work with the moving company to get a better rate. You set up a meeting with the facilities manager to go over the cost estimates for the interior design services and the furniture and fixtures. Perhaps you can cut down on the number of design hours needed and choose cheaper art work for the offices.

Once you get new estimates, you'll revise the budget and then present it to Kate and stakeholders. Once Kate approves and signs off on the budget, this becomes the official cost baseline for the project. You will monitor burn rate, expenditures, and report expenditures and the state of the budget to the stakeholders throughout the remainder of the project.

You set up a meeting with the key stakeholders to determine the list of risks, their probability and impact, response plans, and risk owners, and you record this information in the risk register. Your sample risk register is shown here:

Risk #	Description	Prob	Impact	Risk Score	Response Plan	Risk Owner
1	Moving company availability	.25	1	.25	Y	Procurement manager
2	Bad weather during move	.25	.25	.0625	N	Project manager
3	Furniture order delayed	.10	1	.10	Y	Facilities manager

Summary

Cost estimating is performed after the schedule is created and the resources for the project have been determined. You can use several techniques to create project estimates. Analogous or top-down estimates use expert judgment and historical data to provide a high-level estimate for the entire project, a phase of the project, or a deliverable. Parametric estimating uses a mathematical model to create the estimates and, in its simplest form, multiplies the duration of the project task by the resource rate to determine an estimate. The bottom-up method creates the project estimate by adding up the individual estimates from each work package. Three-point estimates are the average of the most likely, optimistic, and pessimistic estimates.

Cost estimates are used to make up the project budget. The project budget is established by using the organization's chart of accounts and then documenting work effort, duration, equipment and material costs, and any other costs that may be incurred during the course of the project. The cost baseline is the total approved expected cost for the project and is used for forecasting and tracking expenditures throughout the project.

Risk planning involves identifying potential risk events that could occur during the project, determining their probability of occurrence, and determining their impact on the project. Probability is always expressed as a number between 0.0 and 1.0. Risk response plans should be developed for those risks that have a high probability of occurrence, have a significant impact on the project if they occur, or have an overall risk score that is high.

It's important to communicate the risks and response plans to the stakeholders throughout the remainder of the project. If you're working on a project with a long timeline, periodically perform the risk processes again to determine whether your risks are still valid and identify new risks.

Exam Essentials

Know the difference between analogous, parametric, and bottom-up estimating techniques. Analogous, or top-down, estimates use expert judgment and historical data to provide a high-level estimate for the entire project, a phase of the project, or a deliverable. Parametric estimates use a mathematical model to create the estimates. The bottom-up method starts at the lowest level of the WBS and calculates the cost of each item within the work packages to obtain a total cost for the project or deliverable.

Name the two discretionary funding allocations a project may receive. The two types of discretionary funding are a contingency reserve and a management reserve. Contingency reserves are monies set aside to cover the cost of possible adverse events. Management reserves are set aside by upper management and are used to cover future situations that can't be predicted during project planning.

Explain the purpose of a cost baseline. The cost baseline is the total approved, expected cost for the project. It's used in the Executing and Monitoring and Controlling processes to monitor the performance of the project budget throughout the project.

Explain the risk identification process. Risk identification is the process of identifying and documenting the potential risk events that may occur on the project.

Explain the purpose of risk analysis. Risk analysis evaluates the severity of the impact to the project and the probability that the risk will actually occur.

Explain the purpose of risk response planning. Risk response planning is the process of reviewing the list of potential risks impacting the project to determine what, if any, action should be taken and then documenting it in a response plan.

Name the negative risk response strategies. The negative risk response strategies are avoid, transfer, mitigate, and accept.

Name the positive risk response strategies. The positive risk response strategies are exploit, share, enhance, and accept.

Key Terms

Before you take the exam, be certain you are familiar with the following terms:

accept

actual cost (AC)

avoid

bottom-up estimate

budgeting

burn rate

contingency reserve

cost baseline

cost performance index (CPI)

cost variance

earned value (EV)

earned value measurement (EVM)

enhance

estimate to complete (ETC)

expenditure reporting

expenditure tracking

exploit

impact

management reserve

mitigate

order-of-magnitude estimate

planned value (PV)

probability

probability and impact matrix

risk

risk analysis

risk identification

risk planning

risk register

risk response plan

risk response planning

risk triggers

schedule performance index (SPI)

schedule variance

share

SWOT analysis

three-point estimates

top-down estimating

transfer

work effort

Review Questions

1. You are asked to prepare an estimate for a project that involves planting new trees in the parking lot. The trees cost $800 each, and the labor to install them is $75 per hour. You are planting 10 new trees, and each tree takes one hour of labor to plant, stake, and water. What is the estimated cost of the labor for this project, and which technique are you using to determine this estimate? Choose two.

   A. $8,000

   B. Three-point estimate

   C. Bottom-up method

   D. $750

   E. Analogous method

   F. Parametric method

   G. $8,750

2. Top-down estimating is another name for which type of estimating technique?

   A. Parametric estimating

   B. Analogous estimating

   C. Three-point estimating

   D. Expert judgment

3. The total time it will take for one person to complete a task from beginning to end without taking into account holidays, time off, or other project work is known as this.

   A. Duration estimate

   B. Work effort estimate

   C. Bottom-up estimate

   D. Parametric estimate

4. A discretionary fund used by the project manager to cover the cost of possible adverse events during the project is known as which of the following?

   A. Management reserve

   B. Chart of accounts

   C. Contingency fund

   D. Cost baseline

5. You are in the process of determining the cost baseline. All of the following are used in the cost baseline except for which one?

   A. Management reserves

   B. Chart of accounts

   C. Human resource cost estimates

   D. Materials and equipment estimates

6. You are developing a bottom-up estimate for the first phase of your project. Which of the following is the most important input to complete this task?

    A. Historic data from a similar project

    B. Chart of accounts

    C. The WBS

    D. The scope statement

7. What is considered the most accurate estimate?

    A. Analogous estimate

    B. Bottom-up estimate

    C. Estimates based on expert judgment

    D. Parametric estimate

8. You are asked to present and explain your project cost baseline. All of the following are true except which one?

    A. The baseline will be used to track actual spending against the cost estimates.

    B. The baseline can be used to predict future project costs.

    C. The baseline is calculated and approved by the project manager.

    D. The baseline is the total expected cost for the project.

9. Your project task is complex and you decide to use a three-point estimating technique. Which of the following options determine the three-point estimate? Choose three.

    A. Quantity estimate

    B. Work package level estimate

    C. Materials estimate

    D. Pessimistic estimate

    E. Resource estimate

    F. Rate estimate

    G. Optimistic estimate

    H. Most likely estimate

10. How is burn rate typically calculated?

    A. CV

    B. Determining spending rates over time

    C. CPI

    D. AC—PV

11. The work effort multiplied by which of the following will bring about the total estimate for each task?

    A. Duration

    B. Rate

    C. Number of resources

    D. Number of hours

12. Your project has a potential for a future risk event. The sponsor has told you that the organization cannot sustain the consequences of this risk. You recommend purchasing insurance so that if the risk event occurs, the organization can recoup their expenditures for the impacts of the risk. What risk strategy is this known as?

    A. Avoid

    B. Mitigate

    C. Accept

    D. Transfer

13. This technique can be used to help identify risks.

    A. SWOT

    B. CPI

    C. EVM

    D. CV

14. The risk register typically contains several pieces of information. Which of the following would you expect to see on a risk register? Choose three.

    A. Risk owner

    B. Description of risk

    C. Risk score

    D. Cost estimate for response plan

    E. Resource costs to track risks

    F. Cost estimate of the consequences of the risk

15. You have identified a risk on your project, and the team decides they won't create a response plan; if the risk happens, they'll deal with consequences when they occur. This is an example of which risk strategy?

    A. Exploit

    B. Avoid

    C. Mitigate

    D. Accept

16. The difference between planned expenditures and actual expenditures is known as which of the following?

    A. Planned value

    B. Variance

    C. Expenditure reporting

    D. Burn rate

17. The clouds are rolling in over the horizon and the wind is picking up. Your outdoor event is about to get rained out. What is this an example of?

   A. Risk trigger

   B. Risk analysis

   C. Risk probability

   D. Risk response

18. All of the following are strategies for dealing with negative risks, except for which one?

   A. Accept

   B. Transfer

   C. Share

   D. Mitigate

19. Project costs when displayed graphically over time represent which of the following?

   A. S curve

   B. C curve

   C. Evenly distributed expenditures

   D. Erratic expenditures

20. You are determining the risk score for each of the risks in your risk register. You need which of the following to determine this score? Choose two.

   A. Response plans

   B. Risk owners

   C. Probability the risk will occur

   D. Contingency reserves

   E. Risk trigger scores

   F. Impact if the risk occurs

Chapter

8

Communicating the Plan

THE COMPTIA PROJECT+ EXAM TOPICS COVERED IN THIS CHAPTER INCLUDE

✓ **3.1 Given a scenario, use the appropriate communication method.**

- Meetings
  - Kickoff meetings
  - Virtual vs. in-person meetings
  - Scheduled vs. impromptu meetings
  - Closure meetings
- Email
- Fax
- Instant messaging
- Video conferencing
- Voice conferencing
- Face-to-face
- Text message
- Distribution of printed media
- Social media

✓ **3.2 Compare and contrast factors influencing communication methods.**

- Language barriers
- Time zones/geographical factors
- Technological factors
- Cultural differences
- Interorganizational differences

- Intraorganizational differences
- Personal preferences
- Rapport building/relationship building
- Tailor method based on content of message
- Criticality factors
- Specific stakeholder communication requirements
    - Frequency
    - Level of report detail
    - Types of communication
    - Confidentiality constraints
    - Tailor communication style

✓ **3.3 Explain common communication triggers and determine the target audience and rationale.**

- Audits
- Project planning
- Project change
- Risk register updates
- Milestones
- Schedule changes
- Task initiation/completion
- Stakeholder changes
- Gate reviews
- Business continuity response
- Incident response
- Resource changes

Communication is a critical success factor for your project. In this chapter, you'll learn about communication methods, factors that influence communication, and triggers that bring about communication. I'll also discuss managing stakeholder expectations and the types of communication requirements stakeholders have. Lastly, this chapter will wrap up with the topics of obtaining approval and sign-off on the project management plan.

So, let's get started with the most important aspect of any project: communications.

Communications Planning

Good communication is the key to project success. Granted, you need a solid plan including the scope statement, schedule, and budget. But if you aren't able to communicate the plan or keep stakeholder expectations in line with the project goals, you could end up with an unsuccessful project on your hands in spite of having a great plan.

Good communication involves far more than just setting up distribution lists and talking with your stakeholders at the watercooler. You need a plan to determine what gets communicated to whom and when. Communications planning is the process of identifying what people or groups need to receive information regarding your project, what information each group needs, and how the information will be distributed. The communication system should monitor the project status and satisfy the diverse communication needs of the project's stakeholders.

The need for good communication starts from the day the project charter is issued and you are formally named project manager (perhaps even earlier if you've been filling the project manager role informally). As you've already seen, the project charter is the first of many project artifacts that needs to be reviewed with your stakeholders. The scope statement, project schedule, budget, and final project plan are all documents that should be discussed, reviewed, and approved by your stakeholders. But the communication can't stop with reviews and approvals. They'll want to know the status of the schedule and budget, and it will be your job to inform them of potential risk events, changes, and issues that may impact the project. To do that, you need a plan.

You'll start by reviewing some of the general principles of exchanging information.

> **How Much Time?**
>
> According to PMI®, project managers should spend as much as 90 percent of their time communicating.

Exchanging Information

The act of communicating is part of your daily life. Every aspect of your job as a project manager involves communicating with others. Communication is the process of exchanging information, which involves these three elements:

Sender   The *sender* is the person responsible for putting the information together in a clear and concise manner and communicating it to the receiver. The information should be complete and presented in a way that will allow the receiver to correctly understand it. Make your messages relevant to the receiver. Junk mail is annoying, and information that doesn't pertain to the receiver is nothing more than that.

Message   The *message* is the information that is being sent and received. Messages can take many forms, including written, verbal, nonverbal, formal, informal, internal, external, horizontal, and vertical. Horizontal communications are messages sent and received between peers. Vertical communications are messages sent and received between subordinates and executive management.

The message should be appropriate and relevant to the receiver. Information that isn't needed or isn't pertinent to the intended audience is considered noise and will likely be discarded before it's read or heard.

Receiver   The *receiver* is the person for whom the message is intended. They are responsible for understanding the information correctly and making sure they've received all the information.

Keep in mind that receivers filter the information they receive through their knowledge of the subject, cultural influences, language, emotions, attitudes, and geographic locations. The sender should take these filters into consideration when sending messages so that the receiver will clearly understand the message that was sent.

The sender-message-receiver model, also known as the *basic communication model*, is how all communication exchange occurs, no matter what format it takes. The sender encodes the information (typically in written or verbal format) and transmits it, via a message, to the receiver.

Transmitting is the way the information gets from the sender to the receiver. Spoken words, written documentation, memos, email, and voicemail are all transmitting methods.

Decoding is what the receiver does with the information when they get it. They convert it into an understandable format. Usually, this means they read the memo, listen to the speaker, read the book, and so on. The receiver decodes the message by reading it, listening to the speaker, and so on. Both the sender and the receiver have responsibility in this process. The sender must make sure the message is clear and understandable and in a format that the receiver can use. The receiver must make certain they understand what was communicated and ask for clarification where needed.

Project communication always involves more than one person. Communication network models are a way to explain the relationship between the number of people engaged in communicating and the actual number of interactions taking place between participants. For example, if you have five people in a meeting exchanging ideas, there are actually ten lines of communication among all the participants. Figure 8.1 shows a network model showing the *lines of communication* among the members.

FIGURE 8.1   Network communication model

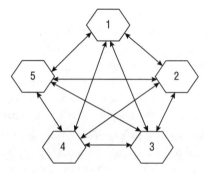

The nodes are the participants, and the lines show the connection between them all. The formula for calculating the lines of communication looks like this:

(Number of participants × (Number of participants –1)) ÷ 2

Here's the calculation in mathematical terms:

$n(n-1) \div 2$

Figure 8.1 shows five participants, so let's plug that into the formula to determine the lines of communication:

$5(5-1) \div 2 = 10$

Effective vs. Efficient Communication

Keep in mind there is a difference between effective and efficient communication. *Effective* communication concerns providing the right information in the right format for the intended audience at the right time. *Efficient* communication refers to providing the appropriate level and amount of information at the right time, that is, only the information that's needed at the time.

Listening

In the sender-message-receiver model, there's one critical communication skill that we all need to possess as a receiver, and that is the art of listening. Listening isn't the part where you plan out what you're going to say because someone else is speaking. It's the part where you actively engage with the sender and ask clarifying questions to make sure you're understanding the message correctly.

You can use several techniques to improve your listening skills. Many books are devoted to this topic, so I'll try to highlight some of the most common techniques here:

- Appearing interested in what the speaker is saying. This will make the speaker feel at ease and will benefit you as well. By acting interested, you become interested and thereby retain more of the information being presented.

- Making eye contact with the speaker is another effective listening tool. This lets the speaker know you are paying attention to what they're saying and are interested.

- Put your speaker at ease by letting them know beforehand that you're interested in what they're going to talk about and that you're looking forward to hearing what they have to say. While they're speaking, nod your head, smile, or make comments when and if appropriate to let the speaker know you understand the message. If you don't understand something and are in the proper setting, ask clarifying questions.

- Another great trick that works well in lots of situations is to recap what the speaker said in your own words and tell it back to them. Start with something like this, "Let me make sure I understand you correctly, you're saying...," and ask the speaker to confirm that you did understand them correctly.

- Interrupting others while they are talking is impolite. Interrupting is a way of telling the speaker that you aren't really listening and you're more interested in telling them what you have to say than listening to them. Interrupting gets the other person off-track, they might forget their point, and it might even make them angry. However, there is a time and a place where an occasional interruption is needed. For example, if you're in a project status meeting and someone wants to take the meeting off course, sometimes the only way to get the meeting back on track is to interrupt them. You can do this politely. Start first by saying the person's name to get their attention. Then let them know that you'd be happy to talk with them about their topic outside of the meeting or add it to the agenda for the next status meeting if it's something everyone needs to hear.

Methods of Communicating

In ancient times, there were a couple of ways of communicating. The primary method was verbal. You talked to your neighbors and others in town to get the latest news. You also made sure you talked to those traveling through from other towns so you could hear the news from other parts of the world.

Written forms of communication were not prevalent until after the invention of the printing press around 1450. Before then, scribes hand wrote information onto papyrus, and before that, pictures were carved and painted onto cave walls. This was an effective form of communicating for those who had access to the cave. But if you couldn't get to the cave, you had to rely on verbal communications.

Today's technology has brought us dozens of communication methods. I'll touch on a few of them after first getting a little deeper into the advantages of written and verbal communications.

Forms of Communicating

The primary forms of communication exchange include verbal and written formats. Verbal communication is easier and less complicated than written, and it's usually faster. The risks with verbal communication are that misunderstandings can take place and there's no record of what was said other than everyone's memories. Written communications are good for detailed instructions or complex messages that people may need to review. The risk with written communication is that stakeholders are inundated with emails, memos, documents, and other information, so your project information could get lost in the sea of information overload.

Deciding whether your information should be verbal or written is an important decision. Most formal project information, such as the project charter, scope statement, budget, and project plan, should be written. You'll most likely have meetings to discuss the contents of these documents, which means the verbal format will be used as well. It's always good practice to think about what you're communicating beforehand and how it will be communicated. Planning your communications before you speak or write is even more critical if your message is sensitive or controversial. Things will go much easier if you send the right message to begin with, rather than apologizing or retracting it later.

Communication Methods

As I said earlier, there are dozens of ways of communicating today. Let's highlight each of the methods found in the CompTIA Project+ exam.

Meetings   Meetings are prevalent during projects. You will have meetings to determine requirements, approve plans, discuss risks, report on project status, and more. Meetings can use both verbal and written forms of communication. For example, you could be sharing the project charter with the meeting attendees, or have printed copies of the budget or schedule to review, or you might provide written status updates as well as discussing the updates at the meeting.

It's good practice to have an agenda for your meetings that highlights the topics that will be covered, who will be presenting, and the time frame for that topic. There are a few meeting types that may show up on the exam. Let's look at each briefly.

Kickoff Meetings   We talked about project kickoff meetings in Chapter 6. These meetings are intended to discuss the project goals, introduce the team members and stakeholders, introduce the project manager, and set expectations for meeting the project timelines and budgets. You may also discuss team member roles and responsibilities at this meeting.

Virtual vs. In-Person Meetings   In-person meetings occur face-to-face. All of the meeting attendees meet in a conference room or an office to discuss the agenda items at the meeting.

Virtual meetings occur when people are not physically present at the meeting. This can be accomplished using video conferencing or voice conferencing. Video conferencing allows the participants to see and hear each other, and depending on the software you're using, you may also be able to share your computer screen so others can see the presentation, slides, spreadsheet, or whatever report you are presenting. Voice conferencing is conducted using a telephone. It is not as effective as video conferencing but works well if team members can't be in the same room at the same time.

Scheduled vs. Impromptu Meetings   Scheduled meetings are put onto the calendar in advance of the meeting. Typically, invites are sent via email with an agenda and the date, time, and location of the meeting. Impromptu meetings typically occur in the moment and are not scheduled ahead of time. This happens on my projects quite often. I will causally ask a team member a question that grows into asking another person or two to join in the discussion, and the next thing you know, we are having an impromptu meeting. Sometimes, these can be a gold mine of inspiration and a brainstorming opportunity. You wouldn't want all of your meetings to be impromptu, but on those rare occasions when they happen naturally, don't be surprised if you find answers to problems and solve some issues that had previously eluded you and the team.

Closure Meetings   Closure meetings occur at the end of the project or at the end of a project phase. These are important because this is where final acceptance of the project deliverables occurs. You will obtain sign-off from the project sponsor that the project is complete and meets expectations.

Email   Email is a written form of communication. It's useful when conveying succinct, easily understood information quickly to many people. Be careful with email because it's easy for the receiver to misunderstand your tone or intention. Email is great for meeting invites, sharing meeting agendas and minutes, and providing a quick update. If you have complex information to convey, it's best to put this into a written report and present the report in person or via a video conference. It's a good idea to send the report ahead of the meeting so everyone has a chance to review it. But the discussion of the contents should occur at a meeting, not in email. And one last tip: before hitting the "reply all" option, please ask yourself twice whether everyone needs to know your response. Most of the time only the sender needs the reply, not everyone on the distribution list.

Fax   The ability to send documents electronically started with the fax machine. These used to be stand-alone machines that scanned each page of your document and sent it over the telephone lines. Today, you can send a fax over the Internet or over the phone line. In most cases, a document original that is signed and then faxed to a bank, a government agency, or a vendor is considered a legally binding document. Contract amendments and change orders are examples of the types of documents you might send to a vendor. Since you have the capability to send multiple page documents, faxing is appropriate for either simple messages or complex messages.

Instant Messaging Instant messaging allows you to communicate with another person via a text message over the Internet. There are many software programs that provide this capability. You can use instant messaging on a computer, tablet, or phone provided the software program you're using has that capability. It's a great mechanism to convey a quick, short message. Instant messaging isn't appropriate for meetings or complex information or for communicating with multiple people at the same time.

Video Conferencing and Voice Conferencing We talked a bit about video and voice conferencing in the meetings section. Video conferencing is an instant communication technique that uses Internet and telecommunications technology to visually display reports, presentations, and/or people's faces. It also provides audio capabilities so you can both see and hear the other people in the conference. Video conferencing is a multiway communication where two or more people participate in the conference. Video conferencing participants can reside anywhere in the world and use their Internet connection and software to connect to the meeting. Some video conferencing requires the use of a telephone for the audio portion of the meeting and a software program to see the desktop of the presenter. Cameras are needed for video conferencing if you intend to visually see the participants.

Voice conferencing also uses telecommunications technology and allows two or more people to participate in a meeting. Voice conferencing can sometimes be awkward if there are many people in the meeting because it's easy to accidentally interrupt or talk while someone else is talking. Like video conferencing, voice conferencing allows participants to join the meeting from anywhere in the world as long as they have access to a phone line.

Face-to-Face Meetings I also discussed face-to-face meetings a little earlier in this section. Face-to-face meetings are where two or more participants are physically present in the same room. These types of meeting are best when delivering complicated material or when you have to discuss negative results or confront a team member to discuss improper behavior or results.

When in Doubt

If you are ever unsure what method of communication to use, default to face-to-face communication. You always have the advantage in face-to-face meetings of not only hearing their words but also reading their body language. If they tense up or look away from you when you're speaking, there is likely something they don't understand or they may not like the message. Conversely, if they are making eye contact, nodding and asking questions, you can be more certain they are understanding the message than you would be if you emailed them or dropped a document off on their desk.

Text Message   A text message is similar to an instant message only text messages are generally conducted using a cell phone. Again, this method should be used for quick, short, easy-to-convey messages and are typically used to communicate with one person, not a group of people. Text messaging isn't appropriate for relaying complex information.

Distribution of Printed Media   There are times when you may provide printed materials as a form of communication. I generally like to print the project charter, the project scope statement, the project schedule, the budget, meeting agendas, status reports, and the risk register. The project charter and scope statement should each be reviewed a few times, usually once in draft form, another when they are believed to be final, and one more time at the approval and sign-off stage. The project schedule, budget, risk register, and meeting agendas are nice to have in printed form at your status meetings. These should be distributed a few days prior to the meeting so participants have the opportunity to review them beforehand. I find two to three days before the meeting is the ideal time to send these types of documents. If you send them out a week ahead or more, they will likely be forgotten and won't be read. Do keep in mind that if you're sending a contract document or one that's dozens of pages or more in length, it is better to send them a week or more in advance because a couple of days isn't enough time to read all the material. Use your judgment on the timing of the distribution based on the complexity of the information.

Documents such as the project status report, meeting minutes, and action items may have a regular distribution schedule. For example, if you hold project status meetings every week, the schedule may call for the distribution of minutes the following day. Distribution schedules should be discussed with your stakeholders. You'll learn about this topic in the next section.

Social Media   Social media is a way to create and share communication to and from your customer base or project team members in an electronic format. It enables interaction, content sharing, and the ability for anyone connected with the project to provide input.

The next questions are, who should get the information, what format should it be in, and when should they get it? You'll look at these topics next.

The Communication Plan

The communication plan can be simple, and you can easily construct a template using a spreadsheet or table format.

You can document an overall communication plan by doing the following:

- Defining who needs information on your project
- Defining the types of information each person or group needs
- Identifying the communications format and method of distribution
- Assigning accountability for delivering the communication
- Determining when the communications will occur and how often

Table 8.1 shows a sample communication plan.

TABLE 8.1   Example communication plan

Stakeholder Name	Communication Type	Format of Communication	Distribution Method	Responsible Person	Frequency
Sponsor	Project status reports, monthly executive status meeting	Written and verbal	Email notification, project repository, face-to-face	Project manager	Weekly status meetings, monthly executive meetings
Sponsor	Schedule, budget, risk updates	Written and verbal	Email notification, special meetings, reports stored in project repository, face-to-face	Project manager	Immediate notice of schedule or budget updates and risk events with scores of medium or high
Stakeholder A	Project status reports, schedule updates	Written and verbal	Email notification, project repository, project meetings	Project manager	Weekly
All stakeholders	Project schedule updates	Written and verbal	Email notification, project repository	Scheduler	Weekly
Stakeholder B	Project status reports	Written and verbal	Email notification, project repository, project meetings	Project manager	Weekly

You may also choose to include information in this plan on how to gather and store information, how to obtain information between communications, and how to update the communication plan.

Although you can use the template from this example to create an overall communication plan for all stakeholder groups, there are some additional considerations when it comes to communicating with your project team that you'll learn about next.

Communicating with Project Team Members

One of your most important jobs as a project manager is communicating with your project team members. It is your responsibility to make sure all the team members understand the project goals and objectives and how their contribution fits into the big picture. Unfortunately, this is an area that is frequently overlooked in communications planning.

Your interactions with your project team will involve both formal and informal communications. *Formal communications* include project kickoff meetings, team status meetings, written status reports, team-building sessions, or other planned sessions that you hold with the team. *Informal communications* include phone calls and emails to and from your team members, conversations in the hallway, and impromptu meetings.

The challenge that project managers face is matching their communication style with that of each team member. Getting input from your team members will help you better communicate with them. If you are scheduling a kickoff meeting or other team-building session, ask for suggestions on agenda items or areas that require team discussion. Team members may have suggestions for the structure and frequency of the team meetings or format for status reporting, based on their previous project experience. The project manager may not be able to accommodate all suggestions, but taking the time to consider input and reviewing the final format will go a long way toward building a cohesive team.

How Much Is Too Much?

I once participated on a project where the project manager created distribution lists for both email and paper documents and sent everything she received that even remotely involved the project to everyone on both lists. She thought she was doing an excellent job of communicating with the team, but the team was going crazy. We were buried with data, and much of it was not relevant to our role on the project. Most of the team members were so overwhelmed with information they stopped reading everything. That, of course, led to the team missing information they actually needed. The project manager did not understand why there was so much confusion among the team members because she had not put any planning into her communications process.

Everyone has a communications method they are most comfortable with. Some of your team members may prefer email, while others prefer phone calls or face-to-face meetings. Some may prefer to drop in on you and share a piece of information they have or get an update from you. For these informal one-on-one types of communications, try to accommodate what is most comfortable for each team member whenever possible.

Managing Stakeholder Expectations and Communication Needs

Managing stakeholder expectations concerns satisfying the needs of the stakeholders by managing communications with them, resolving issues, improving project performance by implementing requested changes, and managing concerns in anticipation of potential problems.

It's the project manager's responsibility to manage stakeholder expectations. By doing so, you will decrease the potential for project failure. Managing the expectations of your stakeholders will also increase the chance of meeting the project goals because issues are resolved in a timely manner and disruptions during the project are limited.

Stakeholders need lots of communication in every form you can provide, and their communication preferences should be documented within the communication plan. If you are actively engaged with your stakeholders and interacting with them, providing project status and resolving issues, your chances of a successful project are much greater than if you don't do these things. Communicating with stakeholders occurs throughout the project.

The CompTIA Project+ exam mentions specific stakeholder requirements. You can document these requirements in your communication plan. Let's take a brief look at each of them.

Frequency   Stakeholders may have differing requirements for project information including the frequency in which they receive the information. Key stakeholders may require more frequent updates than others, so note the frequency in the communication plan.

Level of Report Detail   Like frequency, the level of detail stakeholders require will differ depending on who they are. For example, the project sponsor may need more details on the budget than most other project stakeholders. Stakeholders who represent a specific area of the business will want more detail about the tasks or risks associated with their own areas.

Types of Communication   Be certain to note the types of communication stakeholders need in the communication plan. Some will want every document you produce; others may be interested only in updated status and schedules.

Confidentiality Constraints   You should pay close attention to sensitive or confidential information and who receives it. You wouldn't want to accidentally send a personnel write-up, for example, to everyone on the distribution list. Other confidential information may include contract details, financials, and company-specific information such as trade secrets.

Tailor Communication Style   Stakeholders, like project team members, may have individual preferences for communication styles. Get to know your key stakeholders and ask them what method of communication they prefer, the frequency they want to receive it, and how they want you to update them on general project information versus critical or emergency type information.

Let's take another look at the communication plan with specific stakeholder needs in mind. Table 8.1 listed the following column headers in the communication plan:

- Stakeholder name
- Communication type
- Format of communication
- Distribution method
- Responsible party
- Frequency

You could add additional columns to describe the level of report detail needed and perhaps use a designator such as high, medium, and low to see at a glance if an executive summary level of information will work or if you need to explain items in detail. Table 8.2 shows a sample communication plan with the additional information noted earlier.

TABLE 8.2   Sample expanded communication plan

Stakeholder Name	Communication Type	Format of Communication	Distribution Method	Responsible Party	Frequency	Level of Detail	Confidential
Sponsor	Project status reports, monthly executive status meeting	Written and verbal	Email notification/ project repository	Project manager	Weekly status meetings, monthly executive meetings	H	No
Stakeholder A	Project plan, project status reports	Written	Reports written and stored in project repository	Project manager	Weekly	M	No

Engaging Stakeholders

Remember that a stakeholder is anyone who has a vested interest in the outcome of a project. In Chapter 2, I identified some of the typical project stakeholders: project sponsor, functional managers, customers, project team members, and so on.

On large projects that cross multiple functional areas, you may have stakeholders who are not actively participating in the project or do not fully understand their role. This can occur when large systems are being implemented or new products are being deployed across several geographic regions. For example, if the customer services director does not understand how their team is involved or what the total impact on their group will be, you need to get them connected to what's happening.

It may be useful to develop a stakeholder engagement plan that describes the key points you need to get across.

- Identify which aspects of the project plan to communicate.

- List any known or probable benefits or concerns from the stakeholder.

- Determine the key message to convey to each stakeholder.

Figure 8.2 shows an example of a stakeholder engagement plan using the customer operations scenario for a new product deployment.

FIGURE 8.2   Stakeholder engagement plan

Communications Plan Stakeholder Engagement Example

Project:_____   Stakeholder Group: <u>Customer Operations</u>

	HIGH LEVEL OVERVIEW (5 minutes)	KEY POINTS (30 minutes)	SUPPORTING DETAIL (1-2 hours)
WHY (are we doing this project?)	Expand product offering.	Increase customer base and projected revenue.	Market research
WHAT (does this mean to the stakeholder?)	All sales channels will require training.	Product functionality highlights Training expectations	Product functionality detail Product demos
HOW (will the project goal be achieved?)	Launch product in selected channels on March 7.	Channel sales goals Channel training dates	Sales channel product proficiency
WHEN (will the stakeholder be involved?)	Supply Core Team lead starting November 9.	Development of training Delivery of training	Interface with Human Factors team. Interface with Customer Care.

This particular example uses three scenarios based on the amount of time you are spending with the stakeholder. This approach has two key benefits.

- You have carefully thought out what you need to say.

- You are ready for the next meeting or to extend your current meeting if the stakeholder wants more detail.

The communication plan should be reviewed with your sponsor. If your project requires communication to executive team members, the sponsor can help you by identifying what information the group needs and how and when communication will take place.

Factors That Influence Communications

It's important to keep the lines of communication open with team members so that you're attuned to issues or conflicts that may be brewing. In an ideal world, this is easy to accomplish if your teams are small in number and all the team members have office space right outside your door. You'd know each of them by name and have the opportunity for informal chats to help you all get to know each other better. If that is not possible, make certain to go the extra mile in building relationships with and among your team members so they feel comfortable relating issues and information to you and others on the team.

In today's global world, it's more likely you have team members in various geographical locations. That means several things for you as the project manager. First, you'll have to be aware of time zone issues when scheduling meetings so that some team members are not required to be up in the middle of the night. Due dates may have to be adjusted in some cases to account for the various time zones. Communication preferences and language barriers also come into play. You may have team members who speak different languages. If so, you'll have to determine the best method for communication. In my experience, it's worked well to use two or three forms of communication, especially for critical information, so that there's less chance for misunderstanding.

Cultural differences can have an impact on teams whether they are collocated or dispersed. If you are used to working in the United States, for example, you know that the culture tends to value accomplishments and individualism. U.S. citizens tend to be informal and call each other by their first names, even if they've just met. In some European countries, people tend to be more formal, using surnames instead of first names in a business setting, even when they know each other well. Their communication style is also more formal than in the United States, and although they tend to value individualism, they also value history, hierarchy, and loyalty. In the Japanese culture, most people consider themselves part of a group, not an individual. Japanese people value hard work and success, as most of us do. You should take the time to research the cultural background of your team members and be aware of the customs and practices that will help them succeed and help you in making them feel like part of the team.

Personal Preferences

Regardless of cultural backgrounds, people still have individual preferences for the way they communicate, receive feedback, provide updates, and interact with the team. Personal preferences are as varied as your team members. So again, get to know your team members and their personal preferences in order to gain the most value from your communications.

Technology barriers can have impacts that are unexpected. For example, in some countries, it's not an uncommon occurrence for the electricity to go out for hours at a time or for Internet connections to drop for no reason. These issues can have significant impacts on the project if you're in the midst of a deliverable or troubleshooting a problem. Team members may also have different levels of proficiency with software programs and other technology you're using during the course of the project. Make certain that training is available where needed, or include questions in the interview process about proficiency with the technology used on the project. It wouldn't be a bad idea to add some buffer time to the schedule to account for unforeseen issues with technology.

Remember that the organizational structure itself may also have an impact on the way you manage teams and the way they interact with each other. Intraorganizational differences that exist within your own team can influence the methods you use to communicate. If you have a small team and they are collocated, informal and impromptu meetings may work well.

Interorganizational differences will also influence your communication methods. For example, functional organizations that are hierarchical in nature can have impacts on the team because there are other managers involved in their career and performance evaluation. They also usually direct the work assignments of their team members. This type of interorganizational difference likely calls for formal communications with written reports that are produced on a regular distribution schedule.

As I discussed earlier, use your best judgment when determining the method and content of your communication, and tailor it to your audience. Be certain to take into account everything that could influence the message such as cultural or language differences. Criticality of the message and timeliness of the information are other factors to consider in your communication methods.

 Real World Scenario

The Geographically Dispersed Team

Jim is a senior systems analyst and project manager for a large aerospace contracting firm. He manages aerospace engineers, some of whom live in California, others in Europe and South America, still others in Colorado, and so forth. They design, build, assemble, and deploy rockets. Jim is an expert at managing geographically dispersed teams. Here are some of his tips:

- You have to understand the project thoroughly. All team members have to be clear about what it is you're building. There can be no question about vision.

- When working on large projects, you must break the project into manageable chunks and group similar phases or work efforts together. This assures communication is targeted to the right team members at the right time.

- In the case of geographically dispersed teams, you simply don't have the funding to fly everyone around the country so they can get together to work on the project. Jim and his teams rely heavily on video conferencing, using instant messaging software and voice conferencing to bring people together to discuss drawings, design characteristics, and other components of the project.

- You don't need to be afraid of geographic boundaries when assembling people with the skills you need. A little thinking outside the box might lead to a well-formed, albeit dispersed, team. That being said, you, the project manager, are the one who determines what will and what will not work as communication methods for a given team.

The power of the Internet has greatly impacted the speed with which team members can communicate and bring their projects to fruition.

Communication Triggers

There are several factors that may bring about the need to update stakeholders or otherwise communicate new information to them. These *communication triggers* are common on all projects and are mostly self-explanatory, so I'll spend just a moment or two briefly touching on each.

Audits   Your organization may conduct audits of projects periodically to determine whether they are receiving adequate value for the money spent and/or to determine whether proper processes and procedures are followed. Audits may be conducted from parties inside or outside the organization. This is one of many reasons to have your project plan documented and to save important project artifacts to the project repository.

Project Planning   Communications occurs throughout the project planning process and beyond. As I've discussed throughout the book, anytime you're in the planning stage, whether developing the charter, scope statement, budget, and more, you should be actively communicating with your stakeholders and team members.

Project Change   Anytime there is a scope, budget, or schedule change to the project, it needs to be discussed and communicated with the stakeholders. Chapter 9 will cover change processes in depth.

Risk Register Updates   Risks are more likely to occur and have bigger consequences earlier in the project. As risks occur or new risks are identified, the stakeholders should be informed of the actions taken and the outcomes of the response plans.

Milestones   Milestones can be a communication trigger because it's good news when they are achieved on time and within budget but not so good news when they are not completed on time or within budget. Both cases are cause for an update to the stakeholders.

Schedule Changes   I mentioned changes a little earlier, but schedule changes are so important it's worth mentioning on its own. Schedules have a way of becoming etched in the minds of project sponsors and key stakeholders, and they will continuously remind you of the project end date. They will want to know whether the schedule slips for any reason, and it's generally delightful news to share that you have completed deliverables or milestones ahead of schedule.

Remember that schedules are also one of the triple constraints on any project.

Task Initiation/Completion   This communication trigger is typically at the team level, rather than the stakeholder level. Project team members appreciate hearing "great job" when tasks are completed on time. Starting new tasks is another opportunity for the project team to communicate with the project manager.

Stakeholder Changes   Stakeholders will sometimes request changes to the project. Executive stakeholders are notorious for thinking they don't have to follow process. They also might think because of their status in the organization, any change they request should be granted. If you find yourself in this situation, enlist the help of your project sponsor or other key stakeholders to explain the importance of communicating changes with all the stakeholders and weighing the pros and cons of the change requests.

Gate Reviews   Gate reviews occur at predetermined points in the project. These could be defined by the PMO, or the gate reviews can be designated by the project manager and agreed upon by the stakeholders. Gate reviews may occur at the end of each project phase (Initiating, Planning, Executing, and so on) or once certain milestones or deliverables are achieved.

Business Continuity Response   The business continuity response plan outlines how the business can continue providing its services or products to their customers in case of a disaster. It describes how business will continue operating including where people should report if they can no longer work at their regular location, an emergency contact list, and how to recover from the disaster and resume operations. Obviously, this is a case where communications should be frequent, effective, and efficient.

Incident Response   Incident response is similar to business continuity response only on a smaller, individualized basis. For example, a flood might impact a portion of your warehouse. That would require an incident response to deal with that occurrence. Business can still be conducted, but the flooded parts of the warehouse must be addressed along with any damaged goods. A flood that impacts your day-to-day operations and prevents you from conducting business as usual and/or requires you conduct business operations from another location is an example of a business continuity response.

Project Suspension

In my humble opinion, if your organization has had a disaster and is enacting their business continuity response plan, it's extremely likely your project will be suspended until business operations are restored. The same is true for an incident. Depending on the incident type and impact, you might find the project sponsor putting the project on hold until the incident is resolved and business returns to normal.

Resource Changes   Another communication trigger for your project is resource changes, especially if key resources are impacted. Anytime a project manager changes, there should be a meeting to announce the change. Perhaps you have resources with specialized skills who have taken employment elsewhere or are experiencing personal circumstances that require an extended absence. This should also be communicated to the stakeholders as soon as you can. Resource changes can have many impacts including schedule delays, budget increases, and the introduction of new risks to the project.

 While it's not specifically mentioned in the CompTIA Project+ objectives, you should note that a change to any of the triple constraints (schedule, budget, or scope) is a communication trigger.

If a communication trigger occurs, you'll need to inform the appropriate stakeholders or project team members and perhaps set up a meeting to discuss the implications. Not all communication triggers require notification to every stakeholder. You'll want to determine the target audience, based on the type of information or trigger that's transpired.

 Make certain to target your audience and determine the rationale for notifying stakeholders when a communication trigger occurs so that you are providing the right information to the right stakeholders.

You might consider capturing your triggers in a spreadsheet or similar program so that you know who to notify if a communication trigger occurs, when to notify them, and what method to use. Table 8.3 shows a sample communication trigger plan.

TABLE 8.3   Example communication trigger plan

Trigger	Stakeholder	Reason	Method	Timing of Notification
Project planning complete	All	Obtain sign-off.	Sign-off meeting	At completion of planning phase
Project change	Sponsor and affected stakeholders	Obtain approval/deferral of change.	Project change management meeting	Regularly scheduled change meeting
Risk register updates	Stakeholder A, Stakeholder B, Stakeholder C, Stakeholder D	Update of risk events occurring, response plans put into action, or notification of new risks	Special risk meeting, project meeting	Immediate and at regularly scheduled project meeting
Resource changes	Sponsor, impacted stakeholders	Inform stakeholders of impacts, delays, or risks from resource changes.	Special meetings, project meetings	Immediate and at regularly scheduled project meeting

 Real World Scenario

Main Street Office Move: The Communication Plan

The project plan for the move is coming together. You're working on the communication plan and will obtain stakeholder approval once you meet with Kate. Here is what you know so far.

The following table shows a partial communication plan for the Main Street Office Move project:

Stakeholder Name	Communication Type	Format of Communication	Method	Responsible Party	Frequency	Level of Detail	Confidential
Kate Sponsor	Project status reports, budget, schedule, moving company updates, issues, risk events, final project approval	Written and verbal	In person, project meetings, printed report, email, and phone	Project manager	Weekly	H	No
Jason IT	Project status reports, schedule, risk events, moving company updates	Written and verbal	Project meetings, in person	Project manager, IT team members	Weekly	M	No
Juliette Communication	Project status reports, schedule, risk events, moving company updates	Written and verbal	In person, project meetings, email, or phone	Project manager, scheduler	Biweekly	H	No

There are several communication triggers that will create a need to inform Kate and other stakeholder of updates. You construct a trigger's list and notification process, as shown in the following table:

Trigger	Stakeholder	Reason	Method	Timing of Notification
Project planning complete	All	Obtain sign-off.	Sign-off meeting	At completion of planning phase
Project change	Kate and affected stakeholders	Obtain approval/ deferral of change.	Project meeting	Regularly scheduled change meeting

Trigger	Stakeholder	Reason	Method	Timing of Notification
Risk register updates	Kate, Leah, Jason, Juliette	Update of risk events occurring, response plans put into action, or notification of new risks	Special risk meeting, project meeting	Immediate and at regularly scheduled project meetings
Milestones	All	Completion, delay, or change to milestone	Project meeting, special meeting if negative impacts/ occurrences	At regularly scheduled project meeting
Schedule changes	All	Notification of schedule change	Project meeting, special meeting if negative impacts/ occurrences	Immediate to Kate, at regularly scheduled change meetings
Task initiation/ completion	Stakeholders responsible for task	Notification tasks are completed and make new task assignments.	Department meetings with functional team members	Daily to functional managers, regularly scheduled project meetings
Stakeholder changes	Kate, other stakeholders depending on the type of change	Discuss impacts/ consequences/ benefits of change.	Special meeting	Immediate to Kate, regularly scheduled change meetings, project meetings
Gate review	All	Discuss accomplishments/ next steps.	Gate review meeting, project meeting	Gate review meetings
Resource changes	Kate, impacted stakeholders	Inform stakehold- ers of impacts, delays, or risks from resource changes.	Special meetings, project meetings	Immediate to Kate, special meetings, regularly scheduled project meetings

Summary

Most project managers spend the majority of their time in the act of communicating. Communication is performed using the sender-message-receiver model. Communications planning is a process where you determine who needs what types of communication, when, and in what format, and how that communication will be disseminated. The network communication model shows the lines of communication that exist between any number of project participants. Listening is another important communication skill for any project manager.

There are many methods of communicating including meetings (which can take many forms including in person, virtual, and more), email, fax, video conferencing, voice conferencing, text messages, instant messages, face-to-face, printed media, and social media.

Factors that can influence communication methods include language barriers, time zone or geographical factors, technological factors, cultural differences, interorganizational and intraorganizational differences, and personal preferences. You'll want to build rapport and solid relationships with your team members so that you can overcome any of these factors that may be influencing communication. Stakeholders sometimes have their own needs in regards to communication requirements including frequency of communication, level of report detail, types of communication, and confidentiality constraints. It's also important to tailor your communication style to the preferences of your stakeholders.

Exam Essentials

Describe the importance of communications planning.   Communications planning is the key to project success. It involves determining who needs information, what type, when, in what format, and the frequency of the communication.

Describe meeting types.   Meetings include kickoffs, virtual, in person, scheduled, impromptu, and closure meetings.

Describe communication methods.   Communication methods include meetings, email, fax, instant messaging, video conferencing, voice conferencing, face-to-face, text message, distribution of printed media, and social media.

Describe the factors influencing communication methods.   Language barriers, time zones/geographical factors, technological factors, cultural differences, interorganizational differences, intraorganizational differences, personal preferences, rapport building/relationship building, content of message, criticality factors, and specific stakeholder communication requirements.

Name the common communication triggers on any project.   Audits, project planning, project change, risk register updates, milestones, schedule changes, task initiation/completion, stakeholder changes, gate reviews, business continuity response, incident response, and resource changes.

Key Terms

Before you take the exam, be certain you are familiar with the following terms:

basic communication model

communication triggers

formal communication

informal communication

lines of communication

message

receiver

sender

virtual meetings

Review Questions

1. Why should you spend time developing a solid communication plan? Choose three.
    A. To set performance goals
    B. To set aside time for your own needs
    C. To keep vendors informed
    D. To understand where the blame lies when something goes wrong
    E. To keep stakeholders updated on your progress
    F. To keep team members informed of project progress

2. This person is responsible for understanding the information correctly and making certain they've received all the information.
    A. Sender
    B. Messenger
    C. Project manager
    D. Receiver

3. There are four participants in your upcoming meeting. How many lines of communication are there?
    A. 6
    B. 4
    C. 8
    D. 16

4. You are working on a project that is being implemented in a country different from your country of origin. You also have team members in several locations around the globe. You should consider all of the following specifically in regard to managing teams in this situation except for which one?
    A. Time zones
    B. Cultural differences
    C. Gate reviews
    D. Language barriers

5. All of the following are considered stakeholder communication requirements except for which one?
    A. Frequency
    B. Distribution of printed media
    C. Confidentiality constraints
    D. Tailor communication style to stakeholder needs

6. Your project team is located in differing time zones. You need to hold a kickoff meeting and decide to use which communication method?

    A. Video conferencing

    B. Email

    C. Fax

    D. Voice conferencing

7. Which of the following are considered communication methods?

    A. Email

    B. Meetings

    C. Social media

    D. Distribution of printed media

    E. Text message

    F. Face-to-face

    G. All of the above

    H. A, B, C, E, and F

8. Communications planning is the process of which of the following?

    A. Scheduling a regular meeting for the project team

    B. Developing a distribution list for the stakeholders

    C. Identifying the people or groups that need information on your project

    D. Creating a template to report project status

9. You are a project manager and have a small team who are colocated. Which of the following factors is probably best suited to communicating with your team?

    A. Face-to-face

    B. Personal preferences

    C. Video conferencing

    D. Meetings

10. You are a new project manager working in the PMO. Your project customer is the finance department. The finance department is uncooperative in working with the PMO and following standard project processes. What factor of influence on communication does this represent?

    A. Cultural differences

    B. Personal preferences

    C. Interorganizational differences

    D. Intraorganizational differences

11. You are a project manager for a telecommunications company assigned to a project to deploy a new wireless network using a technology that does not have a proven track record. One of the key stakeholders introduces a change that could impact the schedule. Which of the following does this describe?

    A. Communication method

    B. Factor influencing communication

    C. Communication trigger

    D. Communication preferences

12. The basic communication model consists of which of the following elements? Choose three.

    A. Decoder

    B. Sender

    C. Transmission

    D. Listen

    E. Message

    F. Receiver

    G. Encrypt

13. You have a project employee who had an inappropriate outburst in a meeting. You need to coach this team member and let them know how to better handle a similar situation in the future. What should the project manager's next step be with this employee in this circumstance?

    A. Set up a face-to-face meeting with the employee.

    B. Email the employee and explain this is not appropriate behavior.

    C. Set up a meeting with the functional manager and the employee.

    D. Text message the employee and demand they stop this behavior.

14. All of the following are communication triggers except for which one?

    A. Audits

    B. Incident response

    C. Task initiation

    D. Resource changes

    E. Technological factors

    F. Milestones

15. You engage in hallway conversations, emails, and phone calls with your team members. What is this considered?

    A. Communication triggers

    B. Factors that influence communication

    C. Communication requirement

    D. Informal communication

16. Video conferencing and voice conferencing are examples of this type of meeting.

   A. Impromptu

   B. In person

   C. Virtual

   D. Informal

17. You have an important message to deliver to stakeholders. Which of the following should the project manager do?

   A. Write the message in an email and distribute it to those who need to know.

   B. Tailor the communication method based on the content of the message.

   C. Call an impromptu meeting.

   D. Set up a voice conference meeting.

18. You have a complex message to communicate to the project stakeholders. Which of the following is the best method to use? Choose two.

   A. In-person meeting

   B. Email

   C. Instant message

   D. Voice conference

   E. Informal

   F. Written format

19. Frequency, level of report detail, types of communication, confidentiality constraints, and tailoring your communication style are examples of which of the following? Choose two.

   A. Rapport building/relationship building

   B. Elements of the communication plan

   C. Communication methods

   D. Stakeholder communication requirements

20. Interorganizational differences, personal preferences, rapport building/relationship building, and technological factors are examples of which of the following?

   A. Factors influencing communication methods

   B. Specific stakeholder communication requirements

   C. Communication triggers

   D. Determining the target audience and rationale

Chapter 9

Processing Change Requests and Procurement Documents

THE COMPTIA PROJECT+ EXAM TOPICS COVERED IN THIS CHAPTER INCLUDE

✓ **1.7 Identify the basic aspects of the Agile methodology.**

- Readily adapt to new/changing requirements.
- Iterative approach
- Continuous requirements gathering
- Establish a backlog.
- Burndown charts
- Continuous feedback
- Sprint planning
- Daily standup meetings/Scrum meetings
- Scrum retrospective
- Self-organized and self-directed teams

✓ **3.4 Given a scenario, use the following change control process within the context of a project.**

- Change control process
    - Identify and document.
    - Evaluate impact and justification.
        - Regression plan (reverse changes)
    - Identify approval authority.
    - Obtain approval.
    - Implement change.

- Validate change/quality check.
  - Update documents/audit documents/version control.
  - Communicate throughout as needed.
- Types of common project changes
  - Timeline change
  - Funding change
  - Risk event
  - Requirements change
  - Quality change
  - Resource change
  - Scope change

✓ **3.5 Recognize types of organizational change.**

- Business merger/acquisition
- Business demerger/split
- Business process change
- Internal reorganization
- Relocation
- Outsourcing

✓ **4.3 Identify common partner or vendor-centric documents and their purpose.**

- Request for Information
- Request for Proposal
- Request for Quote
- Mutually binding documents
- Agreements/contract
- Non-disclosure agreement
- Cease-and-desist letter
- Letters of intent
- Statement of work
- Memorandum of understanding
- Service level agreement
- Purchase order
- Warranty

In the course of performing the work of the project, you will continually monitor the results of the work to make certain they meet the specifications of the project plan. Deviations from the project plan can be warnings that changes may be required or have already occurred. Requests for new requirements or changes to the deliverables or scope will surface during the course of the project. Organizational changes can bring about changes to the project as well. A sound change control process will help you and the team deal with these requests effectively, and I'll talk in depth about those processes in this chapter.

You'll also learn about the procurement process along with the types of procurement documents and partner or vendor-centric documents you may need for your project.

Last but not least, you'll learn about the Agile project management methodology. There's a lot to cover in this chapter, so let's get started with a review of the project management plan.

Project Management Plan Review

In previous chapters, I discussed the concept of the cost baseline, the schedule baseline, and the project management plan. You'll recall that the project management plan is the final, approved documented plan that you'll use throughout the remainder of the project to measure project progress and, ultimately, project success. The key components of the project management plan include the following documents:

- Scope statement
- Project schedule
- Communication plan
- Resource plan
- Procurement plan
- Project budget
- Quality management plan
- Risk management plan

This completed plan serves as the baseline for project progress. You'll use this plan during the Executing and Monitoring and Controlling phases of the project to determine whether the project is on track or whether you need to take action to get the work in alignment with the plan. The project management plan is also referred to when a change is requested. It helps determine whether a change has occurred and is also used to determine whether a requested change is aligned with the overall goals and objectives of the project.

The project management plan is also used as a communication tool for the sponsor, stakeholders, and members of the management team to review and gauge progress throughout the project. As such, it's a good idea to obtain sign-off from the sponsor and key stakeholders on the final plan. This helps assure a common understanding of the objectives for the project, the budget, and the timeline and will ideally prevent misunderstandings once the work of the project begins. It will also help serve as a baseline when the change requests start rolling in and stakeholders want "one more little feature" that may or may not be in keeping with the scope of the project. The project management plan should be updated once a change request is approved.

After-the-Fact Plan

I once was involved on a large project where the project management plan was being created as the project work was being performed. This created a scenario where the project manager, the project team members, and other stakeholders did not have a plan to guide execution of the project, nor did they know what could or should be changed as the requests came in.

As you can imagine, confusion was rampant, and to no one's surprise, the project was quickly off-track.

A project management plan is not a reflection of what has already occurred; it is a plan for future work that will meet the goals and objectives of the project. It will be used to determine whether the project was performed within the constraints (time, scope, cost), to determine whether the milestones were completed, and to validate change requests.

Implementing Change Control Systems

Changes come about for many reasons, and most projects experience change during the project life cycle. Some of the common causes of project changes include the following:

- Timeline change
- Funding change
- Risk events
- Requirements change
- Quality change
- Resource change
- Scope change

These change requests might come from the project sponsor, stakeholders, team members, vendors, and others. You'll want to understand the factors that bring about change, such as those listed here, and how a proposed change might impact the project if it's implemented.

The types of changes others may ask for are limitless. In addition, change requests may also take the form of *corrective actions*, *preventive actions*, or *defect repairs*. These usually come about from monitoring the actual project work results. Let's take a look at a brief description of each of these:

Corrective Actions   Corrective actions bring the work of the project into alignment with the project management plan.

Preventive Actions   Preventive actions are implemented to help reduce the probability of a negative risk event.

Defect Repairs   Defect repairs either correct or replace components that are substandard or are malfunctioning.

The most important aspect of change in terms of project management is having a robust change control system in place to deal with the requests. *Change control systems* are documented procedures that describe how the deliverables of the project are controlled, changed, and approved. They also describe and manage the documentation required to request and track the changes and the updates to the project management plan.

The key to avoiding chaos is to manage change in an organized fashion with an integrated change control system that looks at the impact of any change across all aspects of the project plan. Changes, no matter how small, have an impact on the triple constraints (time, cost, or scope), and they may also impact quality or any combination of these factors. Not having a process to analyze the impact of the change and determine whether it's worth the extra time, money, and so on to implement is a recipe for project failure.

In my experience, the three biggest project killers brought about by the project manager are lack of adequate planning, poor risk planning, and inadequate change control processes.

There are several aspects to an effective change management system. You'll look at each of these elements throughout this section. The change management process includes the following:

- Identify and document the change request.
- Track requests in the change request log.
- Evaluate the impact and justification of the change.
- Disposition the request at the change control board (CCB) and approve or deny.

- Implement the change.
- Validate the change and perform a quality check.
- Update the project management plan, update the appropriate project documents, and apply version control.
- Coordinate and communicate with the appropriate stakeholders.

Identify and Document the Change Request

After the project management plan is approved, all change requests must be submitted through the change control system. The processes and procedures for change control should be documented and easily accessible to stakeholders and team members. The process should describe where to obtain a form for a change request, where and who change requests are submitted to for consideration, and a communication process for keeping the requestor apprised of the status.

A change request could be submitted by most anyone working on or associated with the project. Change requests should always be in writing. This means you'll need to devise a template form for stakeholders and others to document the change request, the reason the change is needed, and what will happen if the change is not made. You could include a place for other information on the form that you think will help the review committee determine whether the change should be made. For example, other information could include potential for additional profits, increased marketability, improved efficiencies, improved productivity, social awareness or benefits, greening potential, and so on.

The change request is like a mini-business case that describes the justification, alternatives, and impacts of the change. Typically, the person requesting the change presents the request to the change control board.

Beware! Stakeholders are notorious for asking for changes verbally even though there is a change control process in place. Spend time at the kickoff meeting explaining to everyone where to find the forms and how to follow the process. Make it a point that only change requests that come in via the process will be considered. Verbal requests and drive-bys to the project team will not be accepted.

Change Request Log

A number of things should happen once the change request is submitted. First, it should be assigned an identifying number for tracking purposes. Then, it should be recorded in the change request log. This log is easy to construct in a spreadsheet file. Table 9.1 shows a sample change request log.

TABLE 9.1   Change request log

ID	Date	Description	Requestor	Status	Disposition	Implementation or Close Date
01	11/11	Add a drop-down box on the entry screen.	Nora Smith	Submitted to review committee	Approved	11/13
02	11/14	Implement virtual tape library for backups.	Brett Whatley	For review on 11/25		

You could add other columns to this spreadsheet for tracking purposes, depending on the needs of your project. For example, you might want to add the date of the committee's decision, implementation status, and columns to track costs and hours expended to implement the change.

After the change request is recorded in the tracking log, the next step is an analysis of the change request.

Evaluate the Impact and Justification of Change

The changes are typically evaluated by the subject-matter experts working on the area of the project that the change impacts, along with input from the project manager. The following questions are a good place to start the analysis process:

- Should the change be implemented?
- What's the cost to the project in terms of project constraints: cost, time, scope, and quality?
- Will the benefits gained by making the change increase or decrease the chances of project completion?
- What is the value and effectiveness of this change?
- Is there a potential for increased or decreased risk as a result of this change?

After answering these basic questions, the expert should then analyze the specific elements of the change request, such as additional equipment needs, resource hours, costs, skills or expertise needed to work on the change, quality impacts, and so on. You can use some of the same cost- and resource-estimating techniques we discussed in previous chapters to determine estimates for change requests.

The project manager will also analyze the schedule, the budget, and the resource allocation to determine what impacts will occur as a result of the change. This information is documented (a template comes in handy here as well) and then presented to the review committee, typically called a *change control board*, for approval or rejection.

Evaluating the impacts of the change request may include what CompTIA calls a *regression plan*. You can think of this as the ability to *reverse changes* or back out the changes and revert to the previous state. Information technology projects often have a roll-back plan (or reverse-changes plan). If the change does not perform as expected, you can go back to the previously known, working state.

Keep in mind there is always an opportunity cost involved when analyzing change. When you ask your subject-matter experts to stop working on their tasks in order to examine the impacts of the change request, their work on the project comes to a stop. This trade-off can be difficult to balance sometimes. There isn't a hard and fast rule on this. You'll have to keep an eye on the progress of the project work and the number of requests coming in. I sometimes track the number of hours spent by the team on change request analysis and report this to the stakeholders during our project status meetings. It helps them to be aware of the potential impact to the project if the change requests get out of hand, and if you have a savvy stakeholder group, they'll begin to self-discipline themselves and curtail frivolous requests.

Change Control Board

In many organizations, a change control board is established to review all change requests and approve, deny, or defer the request. CCB members might include stakeholders, managers, project team members, and others who might not have any connection to the project at hand. Some organizations have permanent CCBs that are staffed by full-time employees dedicated to managing change for the entire organization, not just project change. You might want to consider establishing a CCB for your project if the organization does not have one.

A basic CCB assigns approval authority equally among the key stakeholders on the project. Complex projects or projects that involve one business area more than others may require approval authority weighted toward that business unit. For example, let's say you're implementing a new recruitment software program for the human resources department. A change request is submitted that has to do with the recruitment business process and primarily only affects the human resources department. In this case, the human resources stakeholder should have more say in the decision on the change request than other stakeholders.

Typically, the board meets at regularly scheduled intervals. Change requests and the impact analysis are given to the board for review, and they have the authority to approve, deny, or delay the requests. You should note their decision on the change request log shown in Table 9.1.

It's important to establish separate procedures for emergency changes. This should include a description and definition of an emergency, the authority level of the project manager in this situation, and the process for reporting the change after it's implemented. That way, when emergencies arise, the preestablished procedures allow the project manager to implement the change on the spot. This always requires follow-up with the CCB and completion of a formal change request, even though it's after the fact.

In many organizations, the project sponsor is required to approve changes that impact scope, budget, time, or quality if the estimates surpass a certain limit. It's always good practice to inform your sponsor of major changes to scope, budget, time, quality, or any change that has the potential to increase risk. Check with your organization to understand the process for approvals.

Implement the Change

When a change request is approved, you will need to implement the change. This may require scheduling changes, resource changes, or additional funds. Depending on the complexity of the change, you may need to coordinate activities with the project team and schedule the change at an appropriate time. If the change will come at a later stage in the project, make certain to perform the risk processes because a change could introduce the potential for new risks. You'll need to identify the new risks, log them on the risk register, determine their risk score, and develop response plans for them if appropriate.

Validate the Change

Once the change is implemented, you'll need to validate that the change was made and that it met the requirements of the change request. You'll check the quality of the change to assure it was performed accurately and completely. If there are problems, you may need to implement your regression plan and reverse the changes and then evaluate why the change did not function as planned.

Updating the Project Management Plan

Changes to the project will require updates to the affected project documents, including but not limited to the project scope statement, budget, schedule, risk register, and quality plans. (Chapter 10 will cover more about quality.)

Updating the project plan documents is an important step that is sometimes ignored. If your project has an extended timeline, the changes you've made could be long forgotten in the future if they are not documented. Multiple project managers may come and go, and without a record of the change, you could be putting a future project manager, the project, and the organization at risk. Another key reason for updating the project documents is so that you have an accurate record of the project and can use these documents as a starting point on future projects that are similar in nature. You'll also use these documents at the conclusion of the project when you're performing lessons-learned exercises.

Make certain to practice version control when updating project documents. This way, you'll know that you are always working with the latest document. For example, if the schedule has undergone a change previously and now a new request is submitted that will impact the schedule, you'll want to make certain you are updating the latest version of the schedule. Many project management software programs and document repository programs will automatically perform version control for you. If your organization has not implemented this feature, you can accomplish version control in a couple of ways. Include the version number in your document name, such as "Schedule_v1," or the date and time, "Schedule_2018-02-16."

Communicating the Changes

A component of your change process should include a communication plan. Changes should be communicated to the sponsor, stakeholders, and project team members. This can be accomplished in several ways, and one of the most effective is at the project status meeting.

Project status meetings are generally kicked off during the Executing phase to keep stakeholders apprised of progress. Change requests should be a regular agenda item at these meetings. The change request log should also be reviewed to discuss changes that were implemented during the last reporting period and those scheduled for implementation in the next period.

Types of Organizational Change

Scope, schedule, and cost are not the only changes that may impact the project. Organizational changes can have a significant impact on the project. Let's take a look at some of the organizational changes outlined in the CompTIA Project+ exam objectives.

Business Merger/Acquisition   Changes that come in the form of mergers or acquisitions can lead to project changes or sometimes project cancelations. For the exam, it's important to understand the difference between a *merger* and an *acquisition*. A merger is when two businesses come together to perform business as one organization. Once the merger is complete, they are one entity. An acquisition is when one business takes over another. The organization performing the takeover has the power and authority and becomes the decision maker for both organizations. Mergers or acquisitions could change the overall objectives and goals of the project.

Business Demerger/Split   A *demerger* or *split* is the opposite of a merger or acquisition. The organizations that merged or were acquired decide to break into separate entities. A demerger or split is also likely to change the overall goals of the project or perhaps cause its cancellation.

Business Process Change   *Business process changes* typically occur within the organization. An example of a business process change is automating a process that previously was performed on paper. If your project has anything to do with this process or is impacted by the process, changes will be coming your way.

Internal Reorganization   A *reorganization* can impact the resources assigned to and working on your project. This could potentially delay the schedule, especially if you will no longer have the resources originally planned for project activities. You'll need time to negotiate resource availability with the new managers.

Relocation   A *relocation* involves a physical move of the organization, or parts of the organization, and may impact your project. There could be resources (both physical and human) assigned to the project that are targeted for relocation.

Outsourcing   *Outsourcing* occurs when an organization uses external resources to perform business processes and tasks. Outsourcing usually involves hiring outside companies to perform business functions or tasks such as payroll, information technology, security, janitorial services, and so on. Outsourcing project team members, for example, by bringing in consultants to help with the work, can alter the make-up of the team. You should watch for team cohesiveness when there are outside members and make certain to nip conflicts before they arise.

Procurement Planning

Procurement planning is the process of identifying the goods and services required for your project that will be purchased from outside the organization. If your project doesn't require any external resources, you don't need a procurement plan.

One of the first techniques you should use when thinking about the procurement planning process is whether you should make or buy the goods and services needed for the project. *Make-or-buy analysis* determines whether it's more cost-effective to produce the needed resources in-house or to procure them from outside the organization. You should include both direct costs and indirect costs when using the make-or-buy analysis technique. Direct costs are those that are directly attributed to the project, such as costs needed to produce the resource. Indirect costs are those costs associated with overhead, management, and ongoing maintenance costs. You can also use make-or-buy analysis when deciding whether it's more cost-effective to buy equipment or to lease it. Other considerations to take into account in make-or-buy analysis include capacity, skill sets, availability, and company trade secrets.

The procurement process is complex and often involves the legal department. Many organizations have a procurement department that manages all aspects of the process. If that's the case in your organization, you'll want to make certain you understand the forms and processes you're required to follow, or you may end up with some significant schedule delays. Most procurement departments are highly process-driven, and if you miss something along the way, the procurement folks may or may not choose to show mercy and lend a hand in getting the forms through.

Do not underestimate the power your procurement department possesses. I have been involved on projects that were delayed for months because the procurement process was not started in time or I missed one of the myriad forms they require to complete the transaction. Make certain you know and understand your procurement department's rules and processes so that you don't bring about unnecessary time delays on your project.

As the project manager, you're the buyer of goods and services for your project, so I'll cover the procurement process from the buyer's perspective. The organization selling the goods or services is referred to as a vendor, a seller, a supplier, a consultant, or a contractor.

The typical areas where you may need to procure goods or services are discussed next:

Equipment   For some projects, the equipment needs may be fairly simple to determine. If you are developing a new application that requires new hardware, you'll need to obtain the hardware from outside your company. If you're working on a project that requires equipment your organization routinely has available, you'll want to reserve the equipment for the tasks and time frames needed for the project.

Staff Augmentation   Staff augmentation may come about for several reasons. Perhaps your organization lacks the expertise or skills needed in certain areas. Or, there may be other critical projects that have reserved the same resources you need for your project. Projects with a time constraint may also require more resources than are currently available. Contract resources can help fill this gap.

Staff augmentation may range from contracting with a vendor to run the entire project to contracting for specific resources to perform certain tasks. In my experience, staff augmentation is often needed for large, complex projects.

Other Goods and Services   Goods and services that your organization typically does not produce or keep on hand are good candidates for procurement. You may also find that some of the project deliverables are best met by procuring them from outside the organization.

Procurement planning starts with the decision to procure goods or services outside the organization. Once that decision has been made, you need to determine what type of procurement vehicle is best for the purchase you need to make. A simple purchase order may suffice, or you may need a contract.

Statement of Work

If you're working with vendors to perform some or all of the work of the project, it's critical that they know exactly what you are asking them to do. The *statement of work (SOW)* details the goods or services you want to procure. In many respects it's similar to the project scope statement, except that it focuses on the work being procured. It contains the project description, major deliverables, success criteria, assumptions, and constraints. The project scope statement is a good starting point for documenting the SOW.

The project manager should be involved in the process of creating the SOW to ensure accuracy of the project requirements. Vendors use the SOW to determine whether they are capable of producing the deliverables and to determine their interest in bidding on your project work. The SOW must be very clear and precise. Anything in the SOW that is ambiguous could lead to a less-than-satisfactory deliverable.

Many organizations have templates for creating a SOW. This ensures that all required items are covered, and it provides consistent information to vendors. Once the SOW is complete, you're ready to ask for vendors to bid on the work.

Vendor Solicitation

Solicitation is the process of obtaining responses from vendors to complete the project work as documented in the SOW. Typically, a procurement document is prepared to notify prospective sellers of upcoming work. You can prepare the solicitation notice in several ways. The most common are as follows:

Request for Information (RFI)   An RFI is used when you need to gather more information about the goods or services you need to procure. This process will give you a sense for the number of providers or contractors who can provide the goods or services, and you will get an idea of cost. An RFI or RFQ is used when the costs are unknown to you and you need an estimate for the goods or services.

Request for Quotation (RFQ)   RFQ is similar to the RFI. They both serve the same purpose, and most organizations use one or the other of these procurement methods when determining estimates.

Request for Proposal (RFP)   An RFP is submitted when you are ready to procure and begin the work. This process includes submitting the SOW, receiving bids from vendors and suppliers, evaluating the responses, and making a selection.

RFIs and RFQs may be used interchangeably but may also have different meanings in different organizations, so make certain you understand which document to use according to your organization's process. Regardless of what these documents are called, they should include your SOW, information regarding how responses are to be formatted and delivered, and a date by which responses must be submitted. Potential vendors may also be required to make a formal presentation, or they may be asked to submit a bid.

Most procurement processes allow for a meeting with prospective vendors prior to their completing the RFP; this is called a *bidder conference*. This meeting usually occurs right after the RFP is published, and all prospective vendors are invited to the meeting in order to ask questions and clarify issues they may have identified with the RFP. The bidder conference helps assure that vendors prepare responses that address the project requirements.

At the time the procurement documents are distributed (or earlier), you need to develop the criteria the selection committee will use to evaluate the bids, quotes, or proposals you receive.

Vendor Selection Criteria

Most organizations have a procurement department that will assist you with vendor solicitation and selection. They will advise you regarding the information you need to provide and will usually assign a member of their team to manage the vendor selection process and the contract for your project. Some organizations have approved vendor lists made up of vendors that have already met the basic criteria the company requires to do business with them. If that's the case, your solicitation and selection process will be easier because you'll be working with preapproved vendors that have already crossed several of the procurement hurdles required to proceed.

If you're responsible for vendor selection, you'll need to develop criteria to use when evaluating vendor bids or proposals. It helps to decide up front with the sponsor and other key stakeholders who will be involved in the review and selection of vendor proposals. This group should develop the selection criteria as a team and reach agreement ahead of time regarding the weighting of the criteria. These are some of the criteria you should consider when evaluating bids and proposals:

- The vendor's understanding of the needs and requirements of the project
- Cost
- Warranty period
- Technical ability of the vendor to perform the work of the project
- References
- Vendor's experience on projects of similar size and scope
- Vendor's project management approach
- Financial stability of the vendor's company
- Intellectual property rights and proprietary rights

You can use several techniques to evaluate the proposals, including weighted scoring systems, screening systems, seller rating systems, independent estimates, and more. One of the most common methods is using a weighted scoring model or system. The idea is that each of the criteria you're using to evaluate the vendor is assigned a weight. Each vendor is then given a score for each of the evaluation elements, and the weight is multiplied by the score to give an overall score. Table 9.2 shows a sample weighted scoring model using some of the evaluation criteria shown earlier. The scores are assigned a value of 1 to 5, with 5 being the highest score a vendor can earn. You multiply the weight by the score for each element and then sum the totals to come up with an overall score for each vendor. You would almost always choose the vendor with the highest score using this selection method.

TABLE 9.2   Weighted scoring model

Criteria	Weight	Vendor A Score	Vendor A Total	Vendor B Score	Vendor B Total
Understand requirements	5	2	**10**	4	**20**
Cost	3	3	**9**	4	**12**
Experience	4	1	**4**	2	**8**
Financial stability	3	4	**12**	3	**9**
Final weighted score			**35**		**49**

In this example, vendor B has the highest score and is the vendor you should choose.

Sole-Source Documentation

Sometimes you'll have a procurement situation where you can use only one vendor to fulfill your needs. For example, suppose you have a computer system that is unique to your line of business. You've used that software for several years and are ready for an upgrade. You don't want to go through the headache of installing a new system—you simply want to update the one you already have. This situation calls for a sole-source procurement because there's only one vendor that can meet your requirements.

Types of Contracts

A *contract* is a legal, *mutually binding document* that describes the goods or services that will be provided, the costs of the goods or services, and any penalties for noncompliance. Most contracts fall into one of the following categories: fixed-price contracts, cost-reimbursable contracts, and time and materials contracts. Let's take a closer look at each of these here:

Fixed-Price Contract   A *fixed-price contract* states a fixed fee or price for the goods or services provided. This type of contract works best when the product is very well defined and the statement of work is clear and concise. Using a fixed-price contract for a product or service that is not well defined or has never been done before is risky for both the buyer and the seller. This type of contract is the riskiest for the seller. If problems arise during the course of the project and it takes longer to complete a task than anticipated or the goods they were to supply can't be obtained in a timely manner, the seller bears the burden of paying the additional wages needed to complete the task or paying the penalty for not delivering the goods on time.

Cost-Reimbursable Contract   A *cost-reimbursable contract* reimburses the seller for all the allowable costs associated with producing the goods or services outlined in the contract. This type of contract is riskiest for the buyer because the total costs are unknown until the project is completed. The advantage in this type of contract is that the buyer can easily change scope.

Time and Materials Contract   A *time and materials contract* is a cross between fixed-price and cost-reimbursable contracts. The buyer and the seller agree on a unit rate, such as the hourly rate for a programmer, but the total cost is unknown and will depend on the amount of time spent to produce the product or service. This type of contract is often used for staff augmentation, where contract workers are brought on to perform specific tasks on the project.

Partner and Vendor-centric Documents

Organizations may use other types of documents to manage partner relationships or vendor relationships in addition to the documents mentioned earlier. Let's take a brief look at each.

Nondisclosure Agreement   *Nondisclosure agreements (NDA)* are used when organizations engage the services of an outside entity and want to assure that sensitive or trade secret information is not shared outside the organization. An NDA assures that what's discussed, discovered, or developed is kept within the organization.

Cease-and-Desist Letter   A *cease-and-desist letter* informs the other party to stop (cease) doing the activity they're doing and not do it again (desist). A cease-and-desist letter might be sent to someone who is violating copyright laws as an example.

Letter of Intent   A *letter of intent* outlines the intent or actions of both parties before entering into a contract or other mutually binding agreement. It's a negotiable document and can be thought of as an agreement to agree on the terms and conditions.

Memorandum of Understanding (MOU)   A *memorandum of understanding (MOU)* is an agreement that may outline specific performance criteria or other actions between two or more parties. An MOU is used when a legal agreement can't be created between the parties. For example, two or more government agencies may have an MOU that describes the actions, services, or performance criteria between them. Government agencies cannot hold one another accountable legally as you could with a vendor who is not performing per the terms of a contract. Nonperformance by either party under a contract is enforceable in court. An MOU is not legally enforceable.

Service Level Agreement (SLA)   A *service level agreement (SLA)* defines service level performance expectations among two or more parties. For example, the information technology department may have SLAs in place that outline how quickly they will respond to a critical service desk ticket.

Purchase Order (PO)   A *purchaser order (PO)* is typically written by a buyer and describes the specifications and quantities of the goods or services being purchased and the price. Once the PO is accepted by the seller, it is a legally binding document.

Warranty   A *warranty* is usually associated with equipment, materials, software, or supplies. It is a guarantee that the product will meet expectations and perform as stated. A warranty is typically in effect for a specified period of time and expires once the time period is reached.

Agile Project Management

Agile project management is a method of managing projects in small, incremental portions of work that can be easily assigned, easily managed, and completed within a short period of time called an *iteration* or *sprint*. Iterations or sprints (the terms are interchangeable) are always time-bound. Typically, a sprint is a two-week period but can be any short period of time defined and agreed upon by the team. The goal of a sprint is to produce a deliverable, or a tangible portion of a deliverable, by the end of the sprint.

Agile is a highly iterative approach where requirements can be continually defined and refined based on continuous feedback from the product owner.

Agile methodologies have many benefits, and one of the greatest is high stakeholder involvement. It's a flexible approach that is highly interactive. It encourages a great deal of communication between and among the team. The Agile team consists of several members: Scrum master, product owner, stakeholders, team members. Here is a brief description of each.

Scrum Master   The *Scrum master* coordinates the work of the sprint. They also run interference between the team and distractions that might keep them from the work at hand. A Scrum master is a facilitator and helps to educate others in the Agile process. They assist the product owner in maintaining the backlog, prioritizing work, and defining when the work is done. The Scrum master is a facilitator and not a manager. Project team members do not report to the Scrum master.

Product Owner   The *product owner* represents the stakeholders and is the liaison between the stakeholders and the Scrum master. They speak on behalf of the business unit, customer, or the end user of the product and are considered the *voice of the customer*. There should be only one product owner on the team. Communicating with the stakeholders is a critical responsibility of the product owner. They communicate progress and milestones achieved. They determine project scope, schedule, and request the funding needed to complete the work of the project. They manage and prioritize the backlog, which is essentially a list of tasks or work components.

Stakeholders   Stakeholders are people with a vested interest in the project or the outcomes of the project. They interface with the product owner, who informs them of work progress.

Team Members   Team members are responsible for completing backlog items. They sign up for tasks based on the priority of the work and their skill sets. They establish estimates for the work and take on enough tasks to fill the sprint period. Agile teams are self-directed, self-organized, and self-managed.

Sprint Planning

A sprint, which is a time-bound period of work, always starts with a *sprint planning meeting*. During the meeting, team members choose items from the backlog to work on during the sprint. The *backlog* is a list of requirements (both functional and nonfunctional) for the project. The product owner prioritizes the backlog items based on business need, risk, and value to the organization. The Scrum team members break down large requirements into manageable portions of work that can be completed in a sprint.

The backlog is reviewed at the beginning of each sprint, and new requirements can be introduced, changes can be made to existing requirements, or some requirements might be deleted altogether. The product owner typically determines whether a requirement should be added or removed from the backlog. Team members choose which backlog items to work on during the sprint and how much of the work can be accomplished during the sprint. Daily standups (discussed next) are held every day of the sprint.

The sprint items are easily managed with sticky notes on a white board. The backlog items (also known as user stories) are listed in the first column. The next column shows the backlog items that will be worked during the sprint followed by columns noting the progress or stages the work is in. Last there is a column for completed work. Table 9.3 is an example of a Scrum board.

TABLE 9.3   Sample Scrum board

User Stories	Tasks This Sprint	In Progress	In Review	Completed
Story 1	Story 1.1	Story 1.1		
Story 1	Story 1.2		Story 1.2	
Story 2	Story 2.1			Story 2.1
Story 3	Story 3.1	Story 3.1		

Many software programs are available that help manage the backlog items chosen for the upcoming sprint in a more automated fashion. The software program can also create a *burn-down chart*. A burn-down chart shows the remaining work effort (or time remaining) for the sprint. It displays the time period of the sprint on the horizontal axis (usually expressed as days) and the backlog items on the vertical axis (can be expressed as days or hours). At the end of each day, team members update their estimates for the remaining amount of work, which then updates the burn-down chart. Team members and stakeholders can visually see the amount of work remaining in the sprint. Figure 9.1 shows a simplified example of a burn-down chart.

FIGURE 9.1   Sample burn-down chart

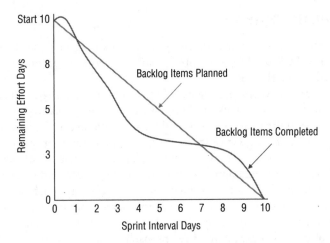

Daily Standups or Scrum Meetings

Daily standups or *Scrum meetings* should be held at the same time and same place every day and should be time limited, usually no more than 15 minutes. Team members must come prepared to discuss the answers to three questions at each meeting:

- What did I accomplish yesterday?
- What will I work on today?
- Do I have any roadblocks or issues preventing me from doing my work?

Standups are an important element in the Agile process. They keep the team informed and alert the Scrum master of any obstacles in the way of completing tasks.

Scrum Retrospective

After the sprint has concluded, a Scrum *retrospective* meeting with the team members, Scrum master, and product owner is held to determine the following:

- Overall progress
- Work that was completed
- Work that was planned but not completed
- Work that needs to carry over into the next sprint
- To review lessons learned to determine how the next sprint, and future sprints, can be improved

If there are backlog items that were not completed during the current sprint, you'll want to discuss this at the next sprint planning meeting. The product owner will use information from this meeting to inform project stakeholders of the overall progress.

Other Methodologies

There are other methods of managing projects you should know about for the exam. Scrum is an original form of Agile used for software development projects. It follows the same processes discussed in this chapter.

Waterfall is an approach where each phase of the project is completed in its entirety before moving to the next phase. This can be a risky methodology because it leaves little room for adding new requirements or functionality, and it doesn't perform reviews or testing of the final product until the end of the project. Final acceptance of the product also occurs at the end of the project after all the deliverables are complete. With Agile or Scrum there are continuous reviews and approvals throughout the project, and you will know right away if you're off-track. You can use an iterative waterfall approach that will reduce risk by completing components of work in a given time frame, but it is still less efficient and effective than the Agile methodology. For the exam, you should know that the waterfall and Agile methodologies each use an iterative approach.

PRINCE2 stands for Projects in Controlled Environments version 2, and it incorporates quality management into the project management processes. It is a comprehensive methodology and is supported by the UK government. It's a methodology that tends to divide projects into multiple stages.

 Real World Scenario

Main Street Office Move

You've implemented a change control process and established a CCB. The first change request came from Jason in IT. After meeting with the facilities manager in the new building, he discovered the power supply feeding the data center will not be adequate. The change request was submitted for $50,000 for additional equipment and services. This will exceed the project budget. You decide to meet with Kate before the CCB meeting to describe the situation and determine whether there is enough contingency funding to cover this cost (should the change request be approved by the CCB) or whether there are available funds elsewhere in the organization to fund this change. You tell Kate that the cost to relocate the fleet cars is coming in below projections, so that is an area that may help fund this change request.

Jason presents the change request at the next CCB, and it is approved. Kate made funding available if the costs exceed the contingency reserve.

You issued an RFP to procure a moving company service to perform the office move. Your selection criteria were as follows: references from recent moves of similar size, delivery dates, and cost. The contract was signed with the moving company after a round of negotiations with them to agree upon delivery dates and times. The moving company responded to the RFP stating they could meet the dates and then wanted to change them during the contract negotiation. You and your procurement officer stuck to the terms of the SOW and the RFP and told them they had to meet the delivery dates stated in the procurement documents or you would find a different moving company.

The project is well underway, and some offices are packed and ready to go. Some fleet cars are already moved to the new garage. You are not using an Agile methodology to manage this project but do find parts of Agile beneficial such as a daily standup meeting with the key stakeholders. A couple of minor issues were brought out at the standup meeting. They were discussed and resolved almost immediately. If they had not been discussed at the standup, the issues could have caused problems or potential delays. You decide to implement standup meetings in future projects during the Executing phase.

Summary

As we discussed earlier in the chapter, things change. Changes come about for many reasons and may take the form of corrective actions, preventive actions, and defect repairs. An integrated change control system manages change requests, determines the global impacts of a change, and updates all impacted portions of the project plan when a change is made. Typically, a change control board is established to review and either approve, deny, or delay change requests.

Integrated change control looks at the overall impact of change and manages updates across all elements of the project plan. Scope change control includes understanding the impact of a scope change, taking appropriate action, and managing a process to review and approve or reject requests for scope changes.

Organizational changes can bring about impacts to your projects as well as the organization itself. Internal reorganizations, mergers and acquisitions, and outsourcing can all bring about changes to the project.

The procurement processes are used when you are purchasing goods or services for the project outside of the organization. You might use an RFI, RFQ, RFP, or PO to procure your goods and services. Sometimes, you may also use resources from other parts of the organization that require formal agreements regarding their use such as an MOU.

Several methodologies are available to manage a project. The Agile methodology involves continuous requirements gathering, is iterative in nature, and is highly interactive. It consists of self-forming and self-directing teams.

Exam Essentials

Describe the project management plan. The project management plan is the final, approved, documented plan that's used in the Executing and Monitoring and Controlling phases to measure project progress.

Describe the elements of a change management process. The elements of a change management process include identifying and documenting the change (using templates and a change log), evaluating the impact, obtaining approval from the CCB, implementing the change, validating the change, updating the project management plan documents, and communicating as needed.

Explain the purpose of a CCB. The change control board reviews, approves, denies, or delays change requests.

Be able to name the types of common project changes. The types of project changes include timeline, funding, risk events, requirements, quality, resource, and scope changes.

Be able to name the types of organizational change. The types of organizational change include business merger, acquisition, demerger, split, business process change, internal reorganization, relocation, and outsourcing.

Be able to describe make-or-buy analysis. Make-or-buy analysis is performed in order to determine the cost-effectiveness of either making or buying the goods and services you need for the project.

Be able to name the types of contracts. The contract types include fixed-price, cost-reimbursable, and time and materials.

Be able to name the types of vendor-centric documents. The types of vendor documents include nondisclosure agreements, cease-and-desist letter, letters of intent, statements of work, memoranda of understanding, service level agreements, purchase orders, and warranties.

Describe the Agile methodology. Agile is an iterative approach to managing projects that readily adapts to new and changing requirements. It provides for continuous requirements gathering and continuous feedback. Agile teams are self-organized and self-directed.

Describe the three primary roles of Agile project management teams. The product owner is the voice of the customer, and they determine the backlog (also known as user stories) and prioritize the backlog. The Scrum master removes obstacles that stand in the way of the team performing its role and provides education on the Agile process. The project team works on backlog items during the sprint and participates in the daily standups.

Name the basic aspects of the Agile methodology. Agile uses daily standups, also called Scrum meetings, to assess progress. Sprint planning occurs at the beginning of each sprint to determine which backlog items to work on. A retrospective meeting is held at the end of the sprint to determine what work was completed and to perform a lessons-learned session on the sprint. Burn-down charts are used to visually display work progress during the sprint.

Key Terms

Before you take the exam, be certain you are familiar with the following terms:

acquisition

Agile project management

backlog

bidder conference

burn-down chart

business process change

cease-and-desist letter

change control board

change control systems

contract

corrective actions

cost-reimbursable contract

daily standups

defect repairs

demerger

fixed-price contract

iteration

letter of intent

make-or-buy analysis

memorandum of understanding (MOU)

merger

nondisclosure agreements (NDA)

outsourcing

preventive actions

PRINCE2

procurement planning

product owner

purchase order (PO)

regression plan

relocation

reorganization

retrospective

reverse changes

Scrum master

Scrum meetings

service level agreement (SLA)

solicitation

split

sprint

sprint planning meeting

statement of work (SOW)

time and materials

voice of the customer

warranty

waterfall

Review Questions

1. You are a project manager for a project developing a new software application. You have just learned that one of your programmers is adding several new features to one of the deliverables. What is the best action to take?

   A. Make any needed adjustments to the schedule and cost baseline, and tell the programmer that any future changes must be approved by you.

   B. Request that the programmer remove the coding for the new features, because he is outside the boundaries of the original scope statement.

   C. Contact the appropriate functional manager, and request a replacement for this programmer.

   D. Determine the source of the request for the new features, and put this change through the change control process to determine the impact of the changes and obtain formal approval to change the scope.

2. Which of the following is not a type of change?

   A. Corrective actions

   B. Defect repairs

   C. Performance corrections

   D. Preventive actions

3. This entity is responsible for reviewing change requests, reviewing the analysis of the impact of the change, and determining whether the change is approved, denied, or deferred.

   A. CAB

   B. CCB

   C. CRB

   D. TRB

4. Which of the following should be established as part of the change control system in the event the change control board (CCB) cannot meet in a timely manner?

   A. Emergency change request procedures

   B. Procedures for analyzing the impacts of change and preestablished criteria for determining which changes can be implemented

   C. Process for documenting the change in the change request log

   D. Coordination and communication with stakeholders

5. After a change request is submitted, all of the following steps occur prior to being reviewed by the change control board except for which one?

   A. The change request is recorded in the change log.

   B. Analysis of the impacts of the change is performed.

    C. Specific elements of the project, such as additional equipment needs, resource hours, quality impacts, and more, are analyzed.

    D. Update the appropriate project planning document to reflect the change.

6. Stakeholders have come to you to tell you they want to change the scope. Before agreeing to the change, what things should you do? Choose two.

    A. Determine which project constraint (time, budget, quality) is most important to stakeholders.

    B. Discuss the proposed scope change with the sponsor.

    C. Ask team members what they think about the scope change.

    D. Define alternatives and trade-offs that you can offer the stakeholders.

    E. Implement the change.

7. You have just received the latest status updates from the team. Based on the progress to date, system testing is projected to take three weeks longer than planned. If this happens, user acceptance testing will have to start three weeks late, and the project will not complete on the planned finish date. The customer scheduled the user acceptance testing participants weeks in advance. What is the best course of action?

    A. Explain to the test team that system test must end on the scheduled date, and they are accountable for the accuracy of the testing results.

    B. Meet with the test team to determine the cause of the delay. If you determine that there are not enough testers to complete all of the scenarios in the time allotted, work with the sponsor to secure additional testers to complete the system test as planned.

    C. Submit the change request to the CCB and, if it's approved, baseline the schedule again.

    D. Escalate the issue of the system test delay to the sponsor, and let them decide what action to take.

8. What is the technique of looking at the trade-offs between producing goods or services internally vs. procuring it from outside the organization?

    A. Cost estimating

    B. Vendor selection criteria

    C. Staff augmentation

    D. Make-or-buy analysis

9. Your project is in danger of being canceled because of an organizational change. Despite the protests of your executive manager, several of the department managers in your old company have been laid off and replaced by the new organization's management team. Which of the following options does this scenario describe?

    A. Your company has experienced a demerger from another organization.

    B. Your company has been merged with another organization.

    C. Your company has been acquired by another organization.

    D. Your company has split from another organization.

10. This document describes the goods or services you want to procure from outside the organization.

   A. RFQ

   B. RFP

   C. RFI

   D. SOW

11. You have just posted an RFP and have invited the vendors to participate in a meeting to ask questions about the work of the project. What is this meeting called?

   A. RFP conference

   B. Bidders conference

   C. Procurement communication conference

   D. Sellers conference

12. This vendor selection method weighs various criteria from the RFP and SOW, scores each vendor on each of the criteria, and determines an overall score for each vendor.

   A. Weighted scoring model

   B. Screening system

   C. Seller rating system

   D. Independent estimates

13. This type of contract is the riskiest for the buyer.

   A. Time and materials

   B. Fixed price

   C. Fixed price plus incentive

   D. Cost reimbursable

14. This type of contract assigns a unit rate for work or goods, but the total cost is unknown.

   A. Time and materials

   B. Fixed price

   C. Fixed price plus incentive

   D. Cost reimbursable

15. You know that the project management plan consists of several project documents and, once approved, serves as the baseline for the project. All of the following are true regarding the project management plan except which one?

   A. It's used during Executing and Monitoring and Controlling phases to determine whether the project is on track.

   B. It's used during the procurement processes to negotiate with the vendor.

   C. It is a communication tool.

   D. It's used when changes are requested to determine whether the change is in keeping with the original goals and objectives of the project.

16. This person is responsible for managing the backlog, prioritizing the backlog, and updating stakeholders with team progress.

    A. Product owner

    B. Scrum master

    C. Sprint owner

    D. Project manager

17. Team members each describe what they accomplished yesterday, what they will work on today, and what obstacles are in their way in this type of meeting. Choose two.

    A. Sprint planning meeting

    B. Standup meeting

    C. Scrum meeting

    D. Project status meeting

    E. Stakeholder meeting

18. These two project management methodologies use an iterative approach. Choose two.

    A. PRINCE2

    B. Six Sigma

    C. Waterfall

    D. Agile

    E. Critical path

19. Your project sponsor has instructed you to use a methodology that allows for continuous feedback and also uses self-organized and self-directed teams. Which methodology does this describe?

    A. PMI®

    B. PRINCE2

    C. Agile

    D. Waterfall

20. What is a backlog? Choose two.

    A. A backlog is the remaining work to be completed in a specific phase of the project.

    B. Backlogs are used in an Agile methodology, and team members choose backlog items at the sprint planning meeting.

    C. A backlog is a list of work that's prioritized by the product owner, but team members choose the backlog items to work on.

    D. Backlogs are used in an Agile methodology, and team members choose backlog items at the sprint retrospective.

    E. Backlog is work not yet completed but was scheduled for a specific iteration.

Chapter

10

Project Tools and Documentation

THE COMPTIA PROJECT+ EXAM TOPICS COVERED IN THIS CHAPTER INCLUDE:

✓ **4.1   Compare and contrast various project management tools.**

- Charts
  - Histogram
  - Fishbone
  - Pareto chart
  - Run chart
  - Scatter diagram
- Dashboard/status report
- Knowledge management tools
  - Intranet sites
  - Internet sites
  - Wiki pages
  - Vendor knowledge bases
  - Collaboration tools
- Performance measurement tools
  - Key performance indicators
  - Key performance parameters
  - Balanced score card

✓ **4.2   Given a scenario, analyze project centric documentation.**

- Issues log
- Status report

- Dashboard information
- Action items
- Meeting agenda/meeting minutes

✓ **1.3 Compare and contrast standard project phases.**

- Closing
    - Transition/integration plan
        - Training
    - Project sign-off
    - Archive project documents.
    - Lessons learned
    - Release resources.
    - Close contracts.

This chapter will look at some of the project tools and documentation needed to inform stakeholders, to document action items and meeting minutes, and to analyze the performance and results of the work of the project.

As the project work winds down, you'll need to review lessons learned, obtain final approval of the project, archive the project documents, and transition the final product of the project to the appropriate business area.

Project managers follow processes to formally close the project. The good news regarding these additional tasks is that much of the work you do during project closure will help you do a better job managing future projects.

Project closure activities apply regardless of the reason the project is ending and regardless of what point you are at in the project life cycle. Even if your project is canceled, there is still a closeout process to perform. One of the elements of project closure involves contract closeout, where there is a formal acceptance (or rejection) of the vendor's work.

Project Management Tools

Project managers have many tools available to help analyze project work and report on progress. According to the CompTIA Project+ objectives, these tools include the following:

- Project scheduling software
- Charts
- Dashboards and status reports
- Knowledge management tools
- Performance measurement tools
- SWOT analysis
- RACI matrix

I've already talked about a few of these such as project scheduling software, SWOT analysis, and the RACI chart. You'll learn about the remaining tools in this section.

Charts

Charts are used in project management to depict the schedule, visually display a process flow, display quality values, and determine performance issues. The charts outlined in the CompTIA Project+ exam include the following:

- Process diagram
- Histogram
- Fishbone
- Pareto chart
- Run chart
- Scatter diagram
- Gantt chart

I covered the process diagram in Chapter 4 and the Gantt chart in Chapter 5. Next up is the histogram.

Histogram

A *histogram* displays the frequency distributions of variable data. It looks like a bar chart, and it's easy to create and understand. The data might include temperature, length, time, mileage, weight, distance, and so on. For example, let's say you are measuring the chemical levels of a certain segment on the production line. Your team takes measurements six times per day for ten days. The ideal range is 2 to 4 parts per million. The histogram in Figure 10.1 shows the frequency of measurements and their distribution.

FIGURE 10.1   Histogram chart

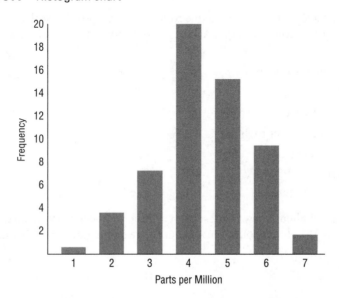

Fishbone

A *fishbone diagram* is a cause-and-effect diagram that shows the relationship between the effects of problems and their causes. This diagram depicts every potential cause of a problem and the effect that each proposed solution will have on the problem. This diagram is also called an Ishikawa diagram after its developer, Kaoru Ishikawa. Brainstorming sessions are a great way to construct a fishbone diagram. The participants can help identify every possible cause of the problem. Let's say your house painting business is experiencing some unhappy customers. They are unsatisfied with the quality of the finished paint job. Some of the causes might be the supplies, the people performing the work, equipment failures, or process problems. Figure 10.2 shows an example cause-and-effect diagram.

FIGURE 10.2   Fishbone diagram

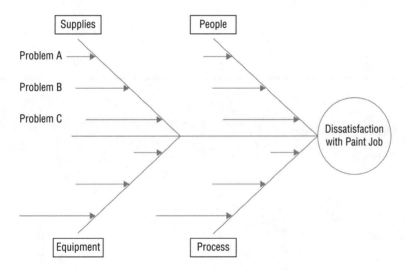

Pareto Chart

You have probably heard of the 80/20 rule. Vilfredo Pareto, an Italian economist and sociologist, is credited with discovering this rule. He observed that 80 percent of the wealth and land ownership in Italy was held by 20 percent of the population. Over the years, others have shown that the 80/20 rule applies across many disciplines and areas. As an example, generally speaking, 80 percent of the deposits of any given financial institution are held by 20 percent of its customer base.

The 80/20 rule as it applies to issues says that a small number of causes (20 percent) create the majority of the problems (80 percent). Have you ever noticed this with your project or department staff? It always seems that just a few people cause the biggest headaches. But I'm getting off-track.

Pareto charts are displayed as histograms that rank-order the most important factors—such as delays, costs, and defects, for example—by their frequency over time. Pareto's theory is that you get the most benefit if you spend the majority of your time fixing the most important problems. The information shown in Table 10.1 is plotted on an example Pareto chart shown in Figure 10.3.

TABLE 10.1 Frequency of failures

Item	Defect Frequency	Percent of Defects	Cumulative Percent
A	800	33	33
B	700	29	62
C	400	17	79
D	300	13	92
E	200	8	100

FIGURE 10.3 Pareto chart

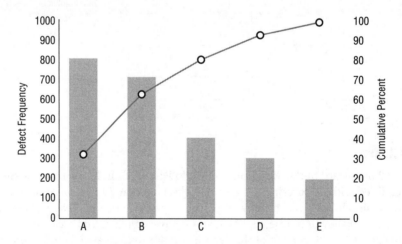

The problems are rank-ordered according to their frequency and percentage of defects. The defect frequencies in this figure appear as black bars, and the cumulative percentages of defects are plotted as circles. The rank-ordering of these problems shows you where corrective action should be taken first. You can see in Figure 10.3 that problem A should receive priority attention because the most benefit will come from fixing this problem.

Run Chart

A *run chart* displays data observed or collected over time as plots on a line. Over time, you will observe patterns and trends such as improvements (or the lack thereof). You can observe positive (or negative) change with a run chart as well. For example, let's say your organization is releasing a new product. They might plot the dollar amount of sales of the new product over a time period. Figure 10.4 shows a sample of this data.

FIGURE 10.4   Run chart

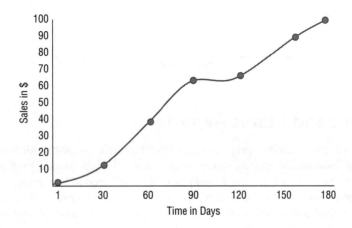

Scatter Diagram

A *scatter diagram* plots two numerical variables on a chart to determine whether there is a correlation between them. Scatter diagrams, also known as correlation charts, display the relationship between these two elements as points on a graph. The closer these variables are to each other, the closer the variables are related. This relationship is typically analyzed to prove or disprove cause-and-effect relationships. As an example, maybe your scatter diagram plots the ability of your employees to perform a certain task. The length of time (in months) they have performed this task is plotted as the independent variable on the x-axis, and the accuracy they achieve in performing this task, which is expressed as a score—the dependent variable—is plotted on the y-axis. The scatter diagram can help you determine whether cause-and-effect (in this case, increased experience over time versus accuracy) can be proved. Scatter diagrams can also help you look for and analyze root causes of problems.

Figure 10.5 shows a sample scatter diagram.

FIGURE 10.5 Scatter diagram

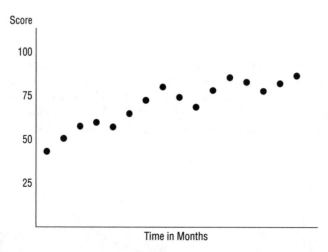

Dashboards and Status Reports

The project manager can analyze project centric documentation in any number of ways. This section will look at dashboards, status reports, and other documents that communicate the status of the project and alert stakeholders with updated information.

Once the project enters the Executing phase, you should conduct regular status meetings that are supplemented with dashboard reports and status reports. *Dashboards* are a way to visually depict the progress of the project and are often used by executives in the organization. They are easy to read and are usually updated with real-time information so they are always up to date. A dashboard typically displays the most important elements of the project such as cost, time, and deliverables. The information is usually succinct and abbreviated. The most common ways to display this data are graphs or charts, numbers, and/or status-level indicators such as Red-Yellow-Green. Red means the project element being reported is in trouble (perhaps over budget, behind schedule, and so on), Yellow means it's moving toward trouble, and Green means everything is going according to plan.

Status meetings should occur regularly during the project. The purpose of the status meeting is to exchange information and provide updated information regarding the progress of the project. You'll want to hold status meetings with the stakeholders, sponsor, project team members, and customers on a regular basis to report on project progress.

 Status meetings may include separate meetings with stakeholders, project team members, and the customer. Take care that you don't overburden yourself with meetings that aren't necessary or meetings that could be combined with other meetings. Having more than three or four status meetings per month is unwieldy.

Send out a *meeting agenda* a day or two before every meeting that describes the topics you and others will address at the meeting. The agenda should state each topic for discussion, the person who will present the topic, and a time frame for each topic. What you report during these meetings depends on the complexity of the project and the makeup of your stakeholder group. At a minimum, you'll want to report on the following items:

- Overall project status
- Schedule updates, including changes
- Milestone achievements
- Budget status
- Change requests this period
- Major issues that could impede progress of the project
- Major risks
- Action items

Meeting minutes should be taken during the meeting so that discussion items and any decisions made are documented. Typically, the project manager hosts the status meetings, and it's difficult to host a meeting and take minutes at the same time. I recommend having someone else on the team take the meeting minutes. Meeting minutes should be distributed to the attendees within a day or two of the meeting and a copy filed with the project documents.

Two other items that are documented and reported on during the status meeting are issues and action items. Let's look at the issues list next.

Issues List

Issues are items that arise that could impact the project, bring about a risk, or delay or prevent the completion of deliverables. Issues should be captured in an *issue log*. The common elements in an issue log include the following:

- Identification number for tracking
- Date the issue was recorded
- Description of the issue
- Name of the person who reported the issue
- Name of the person who owns the issue and will follow through to resolution
- Impact rating of the issue usually expressed as *high-medium-low*
- Action plan to resolve issue
- Status of the issue
- Closed date

Issues should be tracked in the issues log along with their status and resolution. Be aware that an issue log that is not carefully managed can turn into an unwieldy monster, with new issues added weekly and nothing getting resolved.

As you review your issues log with the project team, you want to be sure the person who has been assigned to resolve the issue is working toward closure. Sometimes project issues will remain open for weeks or even months, especially if you consider "we are still working on this one" an acceptable progress report. The status should always include both a plan for resolution and a target date to resolve the issue. If no progress is made, perhaps the responsible party needs assistance or does not really understand the issue. It may also require escalation to the sponsor to overcome roadblocks.

Although the goal is to assign all issues and resolve them as quickly as possible, in some instances you need to prioritize issues and have them worked on in sequence. If multiple issues require the same team member or group, review the impact of each issue on the outcome of the project and establish a priority list.

Action Item

Action items are tasks that come about, usually at a project status meeting, that require follow-up or resolution. They are typically about the project but don't have a direct impact on the work of the project. For example, a stakeholder may ask how much money was spent on the previous work order with the project vendor. This action item is recorded on the action item log, and its results are reported at the next status meeting. Action items can take many forms such as requests for additional information, to prepare a briefing on a specific topic, to contact someone in the procurement department about an upcoming bid, and so on.

 Action items are not a mechanism to create change requests for the work of the project. Changes must follow the change control process.

Action item logs include the following elements:

- Identification number
- Date the action was recorded
- Action item description
- Action item owner
- Progress or resolution to the action item
- Closed date

Status Reports

Status reports are a written summary of the progress of the project and are much more detailed than dashboards. Status reports should be distributed according to the communications plan and a copy retained with the project records. Status reports might take a couple of forms. For example, as the project manager, you might require team members to send you a brief weekly status report that describes the following items:

- Progress since last reporting period
- Progress expected this reporting period

- Progress expected next reporting period
- Obstacles preventing work from being completed

This sounds a lot like the daily standup meetings I talked about in the Agile process. In my experience, I find a daily standup meeting to be more effective for small project teams and more informative than a written status report. If you're managing a large project with several project managers or project assistants with their own teams, a written status report is more efficient and will help you construct the stakeholder status report.

The status report that's distributed to stakeholders should cover several areas of the project. Some of the elements of this status report might include the following:

- Progress since last reporting period
- Progress expected this reporting period
- Progress expected next reporting period
- Review of the project schedule including scheduled and actual completion dates
- Budget review
- Risk review
- Schedule updates or changes
- Resource updates or changes
- Change requests
- Review of issues with high or medium impact
- Notes/comments/announcements

Leave time at the end of the meeting for questions. If you don't know the answers, don't bluff. Let them know you'll research the answer and get back to them and then follow through.

Knowledge Management Tools

Throughout the book I've talked a great deal about communication, project documents, and other artifacts. The amount of documentation on a project can be overwhelming, so I recommend setting up a systematic approach for storing and retrieving all of this information. It's great to have project documents describing the scope, deliverables, schedule, and so on, but if your stakeholders and team members don't know where to find the documents or don't know how to access them, it's as though they don't exist.

The CompTIA Project+ objectives highlight the following knowledge management tools: intranet sites, Internet sites, wiki pages, vendor knowledge bases, and collaboration tools. Let's take a brief look at each.

Intranet and Internet Sites   Intranet sites are internal to the organization and can be accessed only by employees (or authorized personnel) within the organization. They can provide communication mechanisms among and between employees as well as host collaboration sites where project documents can be stored.

Internet sites are external to the organization and are accessed from the intranet via a firewall. Internet sites provide a wealth of information, collaboration tools, and other useful items for managing your project.

Wiki Pages   Wiki pages are typically internal to the organization. Project documents, change control documents, and general information about the project can be posted to the wiki site. A wiki site is generally dedicated to a specific topic or project. Stakeholders, team members, and the project manager should be granted access to the wiki pages to review content. Some team members, depending on their roles, may also have the ability to update documents and post new information. For example, the project coordinator would upload and post project documents as they are approved. The project scheduler would update the project schedule with completed milestones and so on.

Vendor Knowledge Bases   A knowledge base is a collection of information about a product or process that is continually kept up to date. A vendor knowledge base contains information about their products or services. For example, Microsoft has an extensive knowledge base for its products that allows users to query on "how to," various uses of the product, and so on.

Collaboration Tools   Collaboration tools allow project team members to work on the same documents while maintaining version control so only the most recent document is available for review. For example, you might post an SOW and alert your team members that you'd like them to review and make their additions and changes to the document. The collaboration tool provides a check-in and check-out process so that team members are always reviewing and updating the latest version of the document.

Using Performance Measures

The Monitoring and Controlling process group concentrates on monitoring and measuring project performance to identify variances from the project management plan. During this phase, you will collect project data, analyze it, and report on it. The data you'll report on might include information concerning project quality, costs, scope, project schedules, procurement, and risk, and it can be presented in the form of status reports, progress measurements such as key performance indicators, or forecasts.

The project management plan contains the project management baseline data (typically cost, schedule, and scope), which you'll use to monitor and compare performance measurements against. Typically you'll establish performance metrics when developing the plan and then measure them once the work of the project has started. Performance metrics and any deviations from the project management plan should be reported to the stakeholders at the project status meetings.

Key Performance Indicators

A *key performance indicator (KPI)* is a measurable value that shows whether the project is reaching its intended goals. KPIs should be measurable and applicable to the project. For example, perhaps you have KPIs regarding project costs, project management processes, vendor performance, and so on. Almost any element of the project you want to measure can become a KPI. Here are some more specific examples:

- Project costs must not exceed more than 5 percent of the budgeted cost baseline.

- Quality standards will meet or exceed industry standards by no more than +/– .03 standard deviations.

- Increase the number of page visits to the new website by 10 percent over 6 months.

- Increase shared links on social media sites by 15 percent over the next 12 months.

- Increase the number of bookings using self-service check-in to 70 percent over the next two years.

 The important thing to note is that KPIs must be measurable, and they should be communicated to the project team and stakeholders.

If you see that KPIs are not being met or you are finding issues and risk escalating out of control, you need to communicate with your project sponsor. No one likes bad news, but you'll be much better off if you present the facts as soon as you know them. Too many project managers try to ignore problems and hope that the project will turn around—that rarely happens. If you're in doubt, talk with your project sponsor so they can intervene and help before it's too late.

 Real World Scenario

A Phased Delay

A fellow project manager I know had an early project experience that involved a new system application that was being developed for customer care representatives from two recently merged companies. Although the requirements and all of the major deliverables referenced one system, each company had separate back-end systems that needed to interface with the new customer care system. Unfortunately, this piece of information was not discovered until the work of the project started. So, the development team had twice the application interface work to do than was originally planned.

Everyone on the team knew there was no way that the project would be completed as scheduled. The development team provided a revised estimate that showed project completion six months later than the original schedule. Both the development manager and the project manager were afraid to go to the sponsor with this news, so they reported at the next project status meeting there would be a two-week delay and hoped for a miracle. Unfortunately, a miracle didn't occur, so they reported there'd be another two-week delay. At this point, the sponsor started asking a lot of questions, and the project manager had to admit that the best estimates of the additional work indicated a six-month delay. The sponsor was furious that she had not been told the truth from the beginning, and it ruined the credibility of the project manager. In fact, a new project manager was named shortly after this incident.

Key Performance Parameters

Key performance parameters (KPPs) are similar to KPIs, only they are used to set operational goals or performance levels for systems. These are usually represented as the minimum acceptable levels or values for the system. KPPs are used primarily by government entities such as the Department of Defense and the U.S. Army in regard to military systems and equipment.

Balanced Score Card

A *balanced score card* is a strategic management tool used to measure the activities and processes a business uses to meet its strategic goals. It's a way to determine whether the performance of the organization is measuring up to its goals. The balanced score card measures elements such as financial goals, business processes, innovation, the customer experience, and customer satisfaction. Typically, the balanced score card methodology focuses on strategic areas of the business and monitors a small number of important data elements. Balanced score card measures are usually communicated throughout the organization and, in my experience, are also tied to individual performance reviews.

Project Endings

You've made it to the end of the project and to the end of the project lifecycle. The Closing process is the last phase of the project management lifecycle, and it's the most often overlooked. However, there are a few key activities you'll want to complete in the Closing phase. Before diving into those activities, let's look at the characteristics of closing and the reasons projects come to an end.

Characteristics of Closing

A few characteristics are common to all projects during the Closing phase. You've already completed the majority of the work of the project—if not all of the work—so the probability of not finishing the project is low. Risk is low in this process group also because the work is completed. There's little chance that a risk would occur at the end that would derail the project. Stakeholders have the least amount of influence during the Closing processes, while project managers have the greatest amount of influence. Costs are significantly lower during this phase because the majority of the project work and spending has already occurred.

All projects eventually come to an end, and there are several types of project endings I'll cover next.

Types of Project Endings

We all usually think of a project coming to an end when all the deliverables are completed. Ideally, this is what you'll experience most of the time. There are several reasons that a project might end, including but not limited to these:

- They are completed successfully.
- They evolve into ongoing operations and no longer exist as projects.
- Their budgets are slashed.
- The project resources are redirected to other activities or projects.
- The customer goes out of business or is merged with another entity.
- They are canceled prior to completion.

I've worked on many projects that ended in cancelation. This can occur for any number of reasons: project sponsors move on to other assignments, budgets are cut, new management comes into power and changes direction, vendors don't perform as anticipated, and many more. The important thing to remember about cancelation is that all the steps of project closeout should be performed when a project is canceled so that the records are archived and the reasons for cancelation are documented.

All the reasons for project endings I just listed, including cancelation, are incorporated into four formal types of project endings:

- Addition
- Starvation
- Integration
- Extinction

You'll look at each of these ending types in the following sections.

Addition

Projects that evolve into ongoing operations are considered projects that end because of *addition*; in other words, they become their own ongoing business unit or the product or result of the project transitions into an existing business unit before the project is completed.

Starvation

When resources are cut off from the project or are no longer provided to the project, it's starved prior to completing all the requirements, and you're left with an unfinished project on your hands. *Starvation* can happen for any number of reasons:

- Other projects come about and take precedence over the current project, thereby cutting the funding or resources for your project.
- The customer curtails an order.
- The project budget is reduced.
- A key resource quits.

Resource starving can include cutting back or withholding human resources, equipment and supplies, or money. In any case, if you're not getting the people, equipment, or money you need to complete the project, it's going to starve and probably end abruptly.

Integration

Integration occurs when the resources of the project—people, equipment, and supplies—are distributed to other areas in the organization or are assigned to other projects. Perhaps your organization begins to focus on other areas or other projects, and the next thing you know, functional managers come calling to retrieve their resources for other, more important things. Your project will come to an end because of a lack of resources because they have been reassigned to other areas of the business or have been pulled from your project and assigned to another project.

> The difference between starvation and integration is that starvation is the result of staffing, funding, or other resource cuts, while integration is the result of reassignment or redeployment of the resources.

Extinction

This is the best kind of project end because *extinction* means the project has been completed and accepted by the stakeholders. As such, it no longer exists because it had a definite ending date, the goals of the project were achieved, and the project was closed out.

> Sometimes, closing out a project is like finishing a great book. You just don't want it to end. The team is working at peak performance, deliverables are checked off at record pace, and camaraderie is high. If you practice good project management techniques and keep the communication channels open, most of your projects can fall into this category.

Now that you've determined the reason for your project ending, it's time to examine the steps in project closeout and obtain formal written acceptance of the project.

Steps in Closing Out a Project

Closing out a project involves several steps, as shown here:

- Obtaining formal sign-off and acceptance of the project
- Transferring the results of the project to operations and maintenance
- Releasing project resources
- Closing out contracts
- Performing administrative closure
- Documenting historical information for future projects
- Conducting lessons learned
- Preparing the project close report

The project isn't finished until the sponsor or the customer signs on the dotted line. I'll cover obtaining sign-off next.

Obtaining Sign-Off

Project closeout involves accepting the final product or service of the project and then turning it over to the organization. Obtaining formal written sign-off and acceptance of the project is the primary focus of the Closing phase.

Documenting formal acceptance is important because it signals the official closure of the project, and it is your proof that the project was completed satisfactorily. Formal acceptance includes distributing notice of the acceptance of the project results to the stakeholders.

Ideally, obtaining sign-off should just be a formality. If you've involved the sponsor and stakeholders in the verification and acceptance of the deliverables during the Executing and Monitoring and Controlling processes, it should be easy to obtain sign-off on the project.

 The sponsor is the person who has the authority to end the project or accept the final outcome of the project. In cases where you are working on a project that involves an external customer, the sponsor typically is the customer.

Transferring the Product of the Project

Another function of sign-off is that it kicks off the beginning of the warranty period and/or the transfer of the product to maintenance and operations. Sometimes project managers or vendors will warranty their work for a certain time period after completing the project. Projects that produce software programs, for example, might be warranted from bugs for a 60- or 90-day time frame from the date of implementation or the date of acceptance. Typically in the case of software projects, bugs are fixed for free during the warranty period. Watch out, because users will try to squeeze new requirements into the "bug" category mold. If you offer a warranty, it's critical that the warranty spells out exactly what is covered and what is not.

You should document a transition or integration plan for transferring the product or result of the project to the organization. Set up a meeting—or a series of meetings, depending on the complexity of the project—with the manager who will be responsible for the ongoing upkeep of the product or result you're turning over. Provide them with user documentation for the product.

Training is an important component of transition. Document special skills, training, maintenance issues and costs, licensing costs, warranty periods, and so on. Make certain the new manager understands any special requirements for maintaining the product as well.

Releasing Team Members

Releasing team members or other resources may occur once or several times throughout the project. Projects that are divided into phases will likely release team members at the end of each phase. Other times, team members are brought on for one specific activity and are released when that activity is completed. No matter when the team members are released, you'll want to keep the functional managers or other project managers informed as you get closer to project completion so that they have time to adequately plan for the return of their employees. This gives the other managers the ability to start planning activities and scheduling activity dates.

Team members may also become anxious about their status, especially if people are rolling off the project at different times. You should explain to team members that as various deliverables are completed, team members who have completed their assignments are released. Provide your team members with as much information as you can on anticipated release dates.

You should perform a final performance appraisal when releasing team members from the project. If you work in a functional organization, you should coordinate this with the employee's functional manager and make certain your review is included as part of their final, annual review.

Closing Out the Contract

Closing out the contract is the process of completing and settling the terms of the contract and documenting acceptance. This process determines whether the work described in the procurement documentation or contract was completed accurately and satisfactorily.

Procurement documents might have specific terms or conditions for completion and closeout. You should be aware of these terms or conditions so that project closure isn't held up because you missed an important detail. If you are not administering the procurement yourself, be certain to ask your procurement department whether there are any special conditions that you should know about so that your project team doesn't inadvertently delay contract or project closure.

The procurement department needs to provide the vendor formal written notice that the deliverables have been accepted and the contract has been completed. This letter will be based on your approval of the work.

You should retain a copy of the completed contracts to include in the project archives, which I'll discuss next.

Administrative Closure

Administrative closure involves gathering and centralizing project documents, performing a lessons learned review, and writing the final project close report. This is where project records and files are collected and archived, including the project-planning documents, change records and logs, issue logs, lessons learned, and more. You'll also collect and archive documentation showing that the project is completed and that the transfer of the product of the project to the organization (or department responsible for ongoing maintenance and support) has occurred.

Archiving Project Documents

You have created a lot of documents over the course of your project, particularly in the Planning phase. The purpose for archiving those documents is twofold. First, it's to show you have completed the work of the project and that you can produce sign-offs, and other legal documents, should the need arise. The second primary benefit of archiving the project documentation is that it can be used to help you or other project managers on future projects. Your planning documents can be a reference for cost and time estimates or used as templates for planning similar projects in the future.

Check with your project management office (PMO) to determine whether they have a centralized project archive such as an intranet site or wiki pages for the project documents. They will tell you what the guidelines are for documentation and how to file, organize, and store it.

If you don't have a PMO, you'll need to create your own archiving solution. Check with your organization regarding standards compliance and document retention policies. For example, the organization may require all documents to be numbered or named in a certain fashion. Part of your archiving process will include the retention period. Your organization may have guidelines regarding when certain types of documents can be destroyed or what information must be retained. There are also laws regarding retaining some types of documents, so make certain you are familiar with them when creating your archiving site.

Documenting Lessons Learned

A *lessons learned* review session should be conducted at the conclusion of the project or shortly thereafter. The size and complexity of the project will help you decide whether you need to hold one or more review meetings. You'll want to include any key project team members, the project sponsor, and the key stakeholders at a minimum.

The purpose for this review is to assess the good and the not-so-good aspects of the project. During this meeting, you'll evaluate each phase of the project in order to determine the things that went right and the things that could be improved.

> Documenting lessons learned gives you the opportunity to improve the overall quality of your project management processes on the next project and benefits projects currently underway.

Lessons learned describe the successes and failures of the project. As an example, lessons learned document the reasons why specific corrective actions were taken, their outcomes, the causes of performance variances, unplanned risks that occurred, mistakes that were made and could have been avoided, and so on.

Lessons learned help you assess what went wrong and why, not so you can point fingers at the guilty parties, but so that you can improve performance on the next project by avoiding the pitfalls you encountered on this one. It also helps you determine what went right so that you can repeat these processes on the next project. Lessons learned review involves analyzing the strengths and weaknesses of the project management process, the project team, and, if you dare, the project manager's performance.

Unfortunately, sometimes projects do fail. You can learn lessons from failed projects as well as from successful projects, and you should document this information for future reference. Most project managers, however, do not document lessons learned. The reason for this is that employees don't want to admit to making mistakes or learning from mistakes made during the project. And they do not want their name associated with failed projects or even with mishaps on successful projects.

You and your management team will have to work to create an atmosphere of trust and assurance that lessons learned are not reasons for dismissing employees but are learning opportunities that benefit all those associated with the project. Lessons learned allow you to carry knowledge gained on this project to other projects you'll work on going forward. They'll also prevent repeat mistakes in the future if you take the time to review the project documents and lessons learned prior to undertaking your new project.

> Lessons learned can be some of the most valuable information you'll take away from a project. We can all learn from our experiences, and what better way to have even more success on your next project than to review a similar past project's lessons learned document? But lessons learned will be there only if you document them now.

The following is a partial list of the areas you should review in the lesson learned session. This is by no means a complete list but should give you a good starting point.

You should document everything you learn in these sessions. Lessons learned are included with all the other project documentation and go into the project archive when completed.

- Review each process group (Initiating, Planning, Executing, Monitoring and Controlling, and Closing).

- Review the performance of the project team.

- Document vendor performance.

- Examine sponsor and key-stakeholder involvement.

- Review the risks that occurred and the effectiveness of the risk response plans.

- Document risks that occurred that were not identified during the project.

- Evaluate the estimating techniques used for costs and resources.

- Evaluate the project budget vs. actual performance.

- Review the schedule performance, critical path, and schedule control.

- Review the effectiveness of the change management process.

 Real World Scenario

Involving Project Team Members in Lessons Learned

Although you can evaluate the various components of the project on your own using the project management plan and the project results, to get a more comprehensive lessons learned review, you should involve the stakeholders and the team members.

One way to organize a project review session is to make the session interactive. Let the participants know in advance which aspects of the project the review will focus on, and ask them to be prepared to contribute input on both what went well and what did not. You could also distribute some questions ahead of the meeting for them to consider. One question you'll always want to ask is, "If you could change one thing about this project, what would it be?"

You should always set ground rules before you start. You want to stress that the purpose of this session is not to assign blame but to assess the project so that both this team and other project teams can learn from your experience.

Prepare the meeting room in advance with easel paper listing all the areas of the project you want to cover, and provide each team member with a pad of sticky notes. For each topic, ask the team members to post one positive occurrence and one negative. Each negative comment needs a plan for improvement. If they encounter this situation on a future project, what would they do differently?

Requiring a plan for improvement serves two purposes: it engages the team members in the review by making them part of the problem-solving process, and it helps keep those few team members who may only want to whine under control. This is not the time or place to complain.

When you have concluded the session, collect all the notes and use them as input for your written report.

Preparing the Project Close Report

A final project closeout report needs to be prepared and distributed to all the project stakeholders. This is the final status report for the project and should include at least the following:

- Recap of the original goals and objectives of the project
- Statement of project acceptance or rejection (and the reasons for rejection)
- Summary of project costs
- Summary of project schedule
- Lessons learned and historical data

This report is usually prepared after the lessons learned review meeting so that lessons learned and other historical data can be included in the report. You'll distribute this to the stakeholders after they have accepted and signed off on the project.

 Real World Scenario

Main Street Office Move: Project Closure

Since the beginning of this project, you have held regularly scheduled status meetings with the stakeholders and used a mix of standup meetings (every other day) and regular status meetings (once a week) with the project team members. It seemed to work well, but you will use standups on a daily basis on future projects. It seemed that issues raised during the standup meetings were more easily addressed and resolved sooner than issues that waited for a formal review meeting.

Your status meetings addressed the work completed in the previous period, the work expected to be completed in the current period, and a review of issues and action items. You were diligent in keeping the issues log and action item log up to date and current so that by the time the moving day came, all items were resolved. There was an issue that arose during the move itself. You documented this and discussed it at the last status meeting held in the new office.

Once you settle into the new offices, you hold a lessons learned review meeting with the team and stakeholders. At the rate your company is growing, you anticipate another move within three to five years. The lessons learned from this project will help improve the process next time. You also retain the issues log with the lessons learned because some of the issues could be addressed in the next project plan and ideally avoided in the future.

Summary

This chapter covered a lot of ground, starting with charts, a valuable project management tool. Histograms are a type of bar chart that displays data distributed over time. They are easy to construct and understand. Fishbone charts are also called Ishikawa diagrams and are a cause-and-effect diagram. A Pareto chart is a histogram that rank-orders the most important data by their frequency over time. A run chart displays data observed or collected over time as plots on a line. A scatter diagram plots two numerical variables on a chart to determine whether there is a correlation between them. The closer these variables are to each other, the closer the variables are related to each other. This relationship is typically analyzed to prove or disprove cause-and-effect relationships. Scatter diagrams are also known as correlation charts.

Dashboards visually display the status of the most important elements of a project. They are typically used by executives in the organization, as they can see the most up-to-date project information at a glance.

Status meetings and status reports are important communication tools to keep stakeholders informed of project status. Meeting agendas should be sent a few days before the meeting so that participants know what to expect or can read materials ahead of time. Meeting minutes should be distributed shortly after the meeting occurs.

The issue log should be regularly updated to reflect new issues and to document the status of ongoing issues. The issue log should be reviewed at the status meetings. Action item lists should also be updated and reviewed at the status meetings.

Knowledge management tools such as intranet sites, wiki pages, and collaboration tools are useful for creating, updating, storing, and archiving project documents. It's important to keep documents up to date and to also make certain you are working with the latest version of the document.

Key performance indicators (KPIs) are a measurable value that shows whether the project is reaching its intended goals. KPIs should be measurable and applicable to the project. Key performance parameters (KPPs) are measurable values for operational or performance goals associated with systems. Balanced score cards are another strategic management tool used to determine whether the organizational goals are being achieved.

Project closeout should be performed when the project ends or when it's killed or canceled. The Closing process group is the most often skipped on projects because project managers and team members are anxious to move on to their next assignments. It's important to take the time to perform the steps in the Closing process phase so that you can obtain sign-off on the project, turn over the product to the organization, release project resources, close out the contract, document lessons learned, and create a final project report.

Four types of project endings encompass the majority of reasons a project comes to an end. They are addition, starvation, integration, and extinction.

Closing out procurements involves completing and settling the terms of the contract and documenting its acceptance. Product verification occurs here that determines whether the work was completed accurately and satisfactorily.

Administrative closure activities involve gathering and centralizing all the project documents, performing the lessons learned review, and writing the final project report.

Perhaps the most important element of project closure is the lessons learned document. This entails identifying where things went wrong, what things went well, and the alternatives you considered during the course of the project. Lessons learned are an extremely useful reference for future projects regarding what worked and what didn't, for estimating techniques, for establishing templates, and more.

The project close report is distributed to the stakeholders and includes several elements, including the project's goal, the statement of acceptance, a summary of costs and schedule data, and lessons learned data.

Exam Essentials

Be able to explain a histogram.   A histogram displays data distributed over time. It is a type of bar chart.

Be able to explain a fishbone diagram.   A fishbone diagram is a cause-and-effect diagram, also known as an Ishikawa diagram.

Be able to explain a Pareto chart.   A Pareto chart is a histogram that rank-orders data by frequency over time.

Be able to explain a run chart.   A run chart displays data as plots on a timeline.

Be able to explain a scatter diagram.   A scatter diagram displays the relationship between two numerical variables and determines whether they are related to each other. It can also be used to prove or disprove cause-and-effect relationships. Scatter diagrams are also known as correlation charts.

Name the knowledge management tools used for project documents.   The tools include intranet sites, Internet sites, wiki pages, vendor knowledge bases, and collaboration tools.

Name the three performance measurement tools.   They are key performance indicators (KPIs), key performance parameters (KPPs), and balanced score cards.

Be able to describe a status report.   A status report describes the progress of the project to date and usually includes information on scope, cost, and budget.

Name the types of project centric documents.   They include issue log, status report, dashboard information, action items, meeting agenda, and meeting minutes.

Name the four reasons for project endings.   They are addition, starvation, integration, and extinction.

Understand the steps involved in closing a project.   The steps include obtaining sign-off and acceptance, transferring the product to the organization, releasing project resources, closing out contracts, documenting lessons learned, and creating the project closeout report.

Explain the purpose of obtaining formal customer or stakeholder sign-off.   The formal sign-off documents that the customer accepts the project work and that the project meets the defined requirements. It also signals the official closure of the project and the transfer of the final product of the project to the organization.

Describe lessons learned.   Lessons learned describe the successes and failures of the project.

Key Terms

Before you take the exam, be certain you are familiar with the following terms:

action items

addition

balanced score card

dashboards

extinction

fishbone diagram

histogram

integration

issue log

issues

key performance indicators (KPIs)

key performance parameters (KPPs)

lessons learned

meeting agenda

meeting minutes

Pareto charts

run chart

scatter diagram

starvation

status meeting

status reports

Review Questions

1. You want to assure version control of project documents and provide a collaborative platform for team members and stakeholders to review and update certain documents. You could use all of the following knowledge management tools to accomplish this except which one?

   A. Wiki pages

   B. Intranet site

   C. Dashboards

   D. Collaboration software

2. You have just left a meeting with the project sponsor where you were advised that your project has been canceled because of budget cuts. You have called the project team together to fill them in and to review the remaining activities to close out the project. Several of your team members question the benefit of doing a lessons learned review on a project that has been canceled. What should your response be?

   A. Advise the team that part of the review time will be spent on documenting the failure of the lack of clear requirements from the customer.

   B. Tell the team they need to do this to be able to stay on the project payroll another week while they look for a new assignment.

   C. Inform the team that a final report is a requirement from the PMO, regardless of how the project ends.

   D. Explain that there is value both to the team and for future projects in analyzing the phases of the project that have been completed to date to document what went right, what went wrong, and what you would change.

3. You have just left a meeting with the project sponsor where you were advised that your project has been canceled because of budget cuts. You have called the project team together to fill them in and to review the remaining activities to close out the project. Which of the following describes the type of project ending this project experienced?

   A. Extinction

   B. Starvation

   C. Addition

   D. Integration

4. Which of the following measurement tools are used to measure operational or performance goals for systems?

   A. KPIs

   B. KPPs

   C. Balanced score card

   D. Run chart

5. Which of the following is the "best" type of project ending?

   A. Extinction

   B. Addition

   C. Integration

   D. Starvation

6. What is the primary purpose of a formal sign-off at the conclusion of the project work?

   A. The sign-off allows the project manager to start a new assignment.

   B. The sign-off means the project team is no longer accountable for the product of the project.

   C. The sign-off is the trigger for releasing team members back to their functional organization.

   D. The sign-off indicates that the project meets the documented requirements and the customer has accepted the project deliverables.

7. Which of the following charts is a type of histogram?

   A. Scatter diagram

   B. Fishbone

   C. Pareto chart

   D. Run chart

8. What is the focus of the lessons learned report?

   A. The report should cover both the positive and negative aspects of the project, with suggestions for improvement.

   B. The report should primarily summarize the results of the project schedule, the budget, and any approved scope changes.

   C. The report should focus on the project deliverables and any issues that were created by the customer.

   D. The report should cover what went well during the project and should determine which team member or business unit was responsible for failures or issues.

9. Your project is winding down, and some of your team members are anxious about their status. What is the best way to deal with their concerns?

   A. Explain to the team members that they will be released when the project is done.

   B. Let the team members know that you can only discuss their release date with the functional managers.

   C. Establish the same release date for all the team members, even if their work is not completed.

   D. Review the team member release plans with the functional managers. Keep team members and functional managers informed based on the status of the project schedule.

10. You are in the Monitoring and Controlling phase of the project. Several problems have come to light, and you want to know what the causes of the problems are that are generating the effect. You hold a brainstorming meeting with key team members and plot the cause-and-effect scenarios on this type of chart.

  A. Pareto chart

  B. Histogram

  C. Fishbone diagram

  D. Scatter diagram

11. This document records items that usually arise during a status meeting. They concern the project but do not generally impact the project work directly. This document contains a description, an owner, and status, among other items.

  A. Status report

  B. Action items

  C. Issues

  D. Meeting minutes

12. This is used to keep stakeholders and team members up to date. It typically contains information related to budgets, timelines, deliverables, and risks. The issues log and action items are usually distributed with this document.

  A. Status report

  B. Meeting minutes

  C. Wiki pages

  D. Dashboard information

13. Your project evolved over time into an ongoing operation. What type of project ending is this, and what are your next steps? Choose two.

  A. The next step is to write the project close report.

  B. The next step is to inform the project sponsor and stakeholders the project has ended.

  C. The project ending is because of integration.

  D. The project ending is because of addition.

  E. The next step is to perform a postmortem analysis.

14. Which of the following charts are cause-and-effect diagrams or used to determine if there is a cause-and-effect correlation between two numerical variables? Choose two.

  A. Pareto chart

  B. Fishbone

  C. Run chart

  D. Scatter diagram

  E. Histogram

15. This document is produced at the end of the project and reports the final project outcomes.

    A. Lessons learned

    B. Status report

    C. Project close report

    D. Postmortem analysis

16. You're a project manager for a large project. You're in the middle of the Executing phase. The project sponsor has decided to cancel the project because of unexpected cost overruns and resource shortages. What are your next steps? Choose two.

    A. Change vendors to obtain a lower bid for hardware and software components.

    B. Prepare project closure documents.

    C. Perform a lessons learned analysis and release resources.

    D. Ask the sponsor to allow you to redesign the project with fewer deliverables.

    E. Ask the stakeholders to speak to the sponsor.

17. Your project is too slow paced and your executive stakeholders want more up-to-date information available to them in an easy-to-use format. They are monitoring the project to determine whether it should be canceled or further actions taken to get it back on track. Which of the following is the best option for you to implement?

    A. Dashboards

    B. More frequent status reports

    C. A daily email with status

    D. Invite the stakeholders to the daily stand up meetings so they can hear up-to-date status for themselves.

18. All of the following are true regarding the release of team members except for which one?

    A. Team members are released after lessons learned are documented.

    B. The project manager should perform a final performance appraisal for team members when they're released from the project.

    C. The project manager should inform the functional managers well in advance of the team members' release date.

    D. The project manager should communicate with the team members about their upcoming release date.

19. Your project has experienced some setbacks, and your stakeholders are not happy with progress. A hurricane wiped out one of your vendor's warehouses, and you are scrambling for parts. You work with the vendor and your procurement department to find other suppliers who may have the parts you need. Which of the following is the best option given this scenario?

    A. Cancel the project.

    B. Document the situation in the meeting minutes.

    C. This is an issue that should be recorded and tracked in the issues log.

    D. This is a KPI that the vendor has not met.

20. Who is responsible for authorizing the closure of the project?

    A. Stakeholders

    B. Project manager

    C. Executive team members

    D. Sponsor

Appendix

Answers to Review Questions

Chapter 1: Initiating the Project

1. C, E. A project creates a unique product, service, or result and has defined start and finish dates. Projects must have resources in order to bring about their results, and they must meet the quality standards outlined in the project plan. Interrelated activities are not projects because they don't meet the criteria for a project. Project management processes are a means to manage projects, and processes used to generate profits or increase market share do not fit the definition of a project. Processes are typically ongoing; projects start and stop.

2. B. The Project Management Institute (PMI®) is the leading professional project management association, with more than 700,000 members worldwide.

3. D. A program is a group of related projects that can benefit from coordinated management. Life cycles are the various stages a project goes through, and process groups consist of Initiating, Planning, Executing, Monitoring and Controlling, and Closing.

4. A, C. Portfolios consist of programs, subportfolios, and independent projects that are not necessarily related to one another. An organization could have any number of portfolios.

5. B. Project managers can spend up to 90 percent of their time communicating. The other skills listed here are important as well, but the clue in this question is the 90 percent figure that relates to the amount of time project managers may spend communicating.

6. A. A request to develop a product for use by an internal department is a business need. Market demands are driven by the needs of the market, legal requirements come about because of rules or regulations that must be complied with, and technological advances are because of improvements in expertise or equipment.

7. B, E. A matrix organization can be structured as a strong, weak, or balanced matrix. Employees are assigned to projects by their functional managers, and the project tasks are assigned to them by the project manager. The project manager has the majority of power in a projectized organization.

8. A. A projectized organization is designed around project work, and project managers have the most authority in this type of structure. Project managers have the least amount of authority in a functional organization; they have some authority in a balanced matrix and a little more authority in a strong matrix, but not as much authority as they have in a project-based organization.

9. A. The discounted cash flow technique compares the total value of each year's expected cash inflow to today's dollar. IRR calculates the internal rate of return, NPV determines the net present value, and cost-benefit analysis determines the cost of the project versus the benefits received.

10. D, E. The steps required to validate a project are validating the business case (which encompasses a feasibility analysis, justification for the project, and alignment to the strategic plan) and identifying and analyzing stakeholders.

11. C. Negotiating involves obtaining mutually acceptable agreements with individuals or groups. Leadership involves imparting a vision and motivating others to achieve the goal. Problem-solving involves working together to reach a solution. Communicating involves exchanging information.

12. A, B, C, E. The business case establishes the justification for the project, how it aligns to the strategic goals of the organization, the business need or opportunity that brought about the project, alternative recommendations and analysis, a recommendation on which alternative to choose, and the feasibility study or the feasibility study results may or may not be included in the business case.

13. D. Payback period is a technique that calculates the expected cash inflows over time to determine how many periods it will take to recover the original investment. IRR calculates the internal rate of return, NPV determines the net present value, and discounted cash flows determine the amount of the cash flows in today's dollars.

14. D. The next best step to take in this situation is to perform a feasibility study. Feasibility studies are typically undertaken for projects that are risky, projects that are new to the organization, or projects that are highly complex. Projects of significant risk to the organization shouldn't be taken to the selection committee without having a feasibility study first, and writing the project plan doesn't make sense at this point because you don't know if the project will be chosen or not. You also can't reject the project because there isn't enough information to determine whether it should be rejected until the feasibility study is completed.

15. C. The project manager is ultimately responsible for managing the work of the project. That doesn't mean they should work without the benefit of input from others.

16. D. The key problem with a projectized organization is that there may not be a new project in place at the conclusion of the one team members were released from. This leaves specialists "sitting on the bench" with no work to do and is costly to the organization. It's an advantage to a projectized organization to work on projects. Costs aren't necessarily any higher in this type of organization than others. Costs will depend on the type of project you're working on, not the organizational structure. And the project managers have control over who works on the projects in a project-based organization.

17. B, D, F. The needs or demands that bring about a project include the following: market demand, strategic opportunity/business need, customer request, technological advances, legal requirements, environmental considerations, and social needs. A feasibility study is conducted to determine the viability of a project, and the business case documents the reasons for the project and the justification for the project. Stakeholder needs may bring about a project, but their needs will fall more specifically into one of the seven needs or demands that bring about a project.

18. C. NPV is calculated by subtracting the total of the expected cash inflows stated in today's dollars from the initial investment. In this question, the initial investment is higher than the cash flows, so the resulting NPV is less than zero, and the project should be rejected. Discounted cash flows tell you the value of the cash flows in today's dollars.

19. B. This question describes the expert judgment form of project selection. The question states the executives already read the business case analysis. The feasibility study is a study conducted to determine the risks and potential benefits to the project, and decision models are mathematical models that use differing variables to determine a decision.

20. C, D. The business case analysis may include the feasibility study but should always include the justification for the project and the alignment to the strategic plan. It's a good idea to also include high-level timelines and estimated budgets.

Chapter 2: Project Team Roles and Responsibilities

1. B, C. A stakeholder matrix includes the stakeholder name, department, contact information, role on the project, needs, concerns, interests, level of involvement on the project, level of influence over the project, and notes for your own reference. A stakeholder matrix is an artifact.

2. A. The project sponsor authorizes the project to begin and approves and signs the project charter.

3. C. The project coordinator assists the project manager with administrative functions on the project.

4. B. A key responsibility of the project manager is informing the sponsor of changes, status, issues, and conflicts on the project. The project requestor and stakeholders should be informed as well, but an important aspect of the project manager's role involves informing the sponsor and keeping them updated.

5. C, D, F. The PMO provides standards and practices for the organization including tools, templates, and governance processes.

6. B. The functional manager provides and assigns employees to work on the project. The project manager is accountable for overseeing the work required to complete the project. The customer is the person or group that is the recipient of the product or service created by the project. The project sponsor champions the project throughout the organization and acts as an advisor to the project manager.

7. B, C, E. The project sponsor/champion approves funding, approves the project charter, markets the project benefits, removes roadblocks, and defines the business justification for the project.

8. D. The high-level scope definition describes the reason for the project, its objectives, and the high-level deliverables.

9. C. The project coordinator is responsible for supporting the project manager, providing cross-functional coordination, documentation, time and resource scheduling, and checking for quality.

10. A. You must clarify the project request to determine exactly what the marketing person needs. You need to understand the problem that needs to be addressed so that you can define the high-level requirements and write the project charter.

11. B. The project champion is someone who understands the goals of the project and serves as a voice of enthusiasm throughout the organization regarding the benefits of the project.

12. E. The project sponsor is an individual who is authorized to provide funding and resources necessary to bring about the deliverables of the project. In this question, there isn't enough information to make a good decision about identifying the project sponsor.

13. D. The project manager manages quality assurance, scope, risk, budget, and time, and is also responsible for artifacts.

14. D. The project customer, also known as a client, is the recipient of the product or service of the project.

15. A, E, F. Project team members contribute expertise to the project, contribute deliverables according to the schedule, estimate task durations, and estimate costs and dependencies.

16. A. Project schedulers are responsible for developing and maintaining the project schedule, communicating the timeline and changes to the timeline, reporting on schedule performance, and obtaining task status from resources.

17. A. The project manager should take the time to define the problem or need generating the project request and document this in the high-level scope definition. It could be that the business case was not well defined, but the reason for that is also because the problem or need was not well defined.

18. D. The PMO establishes governance processes for projects as well as providing tools, templates, and more for managing projects.

19. C. The project description describes the characteristics of the product, service, or result of the project. The project description is documented as part of the process of defining high-level scope.

20. B, F, G. The project baseline includes the approved schedule, cost, scope, and quality plans and documents and is approved by the project sponsor. The project baseline is then used to measure performance as the project progresses. You can refer back to the project baseline at any time to determine whether you are on schedule, within scope, and within budget, and to determine whether the quality standards are on target.

Chapter 3: Creating the Project Charter

1. C. The Initiating process concerns the formal acceptance of the project and authorizes the project manager to start the project work. Assigning work to project team members, sequencing project activities, and coordinating resources occur in the Planning process.

2. A. The project sponsor authorizes the project to begin and approves and signs the project charter.

3. D. Monitoring and Controlling deals with risks, performance measuring and reporting, quality assurance, governance processes, change control, and budget control.

4. B. A key role of the project manager is informing the sponsor of changes, status, issues, and conflicts on the project. The project requestor and stakeholders should be informed as well, but the primary role of the project manager involves informing the sponsor and keeping them updated.

5. B, D, F. The activities in the Planning process include the following: project schedule, work breakdown structure, resources, detailed risks, requirements, communication plan, procurement plan, change management plan, and budget. The project charter does include a high-level budget but the project budget is fully developed in the Planning process and is continually monitored throughout the remainder of the project.

6. A, B, E. The Closing process is where the integration plan and transition of the product of the project to other areas of the organization is performed. Lessons learned, project sign-off, close contracts, releasing resources, and archiving documents occur during this process as well.

7. B, C, F. The five process groups are Initiating, Planning, Executing, Monitoring and Controlling, and Closing. Options A, D, and E are Knowledge Areas.

8. D. The Executing process is where the work of the project is performed.

9. C. The project charter formally approves the project and authorizes work to begin. The project schedule and cost estimates are developed later in the Planning process.

10. A. After the project charter is signed and approved, you should hold a kickoff meeting with key stakeholders and key team members to discuss the goals of the project.

11. C. Milestones are major events in a project that are used to measure progress. They may also mark when key deliverables are completed and approved. Milestones are also used as checkpoints during the project to determine whether the project is on time and on schedule.

12. E. All of the options listed are true. If the project charter is signed, you have completed the Initiating process, and the sponsor has officially approved the funds and resources for the project. If the team is anxious to start working right away, they are jumping ahead to the Executing process group.

13. C, D, E. The project charter does not include a high-level cost-benefit analysis or the business case. The business case is its own document and is not part of the project charter. The business case is where the cost-benefit analysis is documented.

14. A, D. Deliverables are an output or result that must be completed in order to consider the project complete. Milestones are used to measure performance.

15. D, E, F. The Initiating phase produces the project charter, business case, high-level scope definition, and high-level risks.

16. B. Assumptions are things believed to be true. In this case, you have not verified Randy's availability and are assuming the functional manager will agree to assign him to the project.

17. B, C, F. Deliverables are measurable outcomes or results or are specific items that must be produced in order to consider the project complete. Deliverables are tangible and are easily measured and verified. Requirements provide detailed characteristics of the deliverables.

18. D. Constraints restrict or dictate the actions of the project team and may take the form of budget, resources, schedules, or other limitations. Situations believed to be true are assumptions.

19. C. The project description describes the characteristics of the product, service, or result of the project.

20. A. Risks are potential future events that pose either opportunities or threats to the project. This is a potential event that would have negative consequences to the project if it were to occur.

Chapter 4: Creating the Work Breakdown Structure

1. C. The key components of scope planning are the scope management plan, scope statement, and work breakdown structure. The project charter is created during project initiation.

2. A. The scope statement serves as a basis for understanding the work of the project and for future decision-making.

3. D. A WBS is a deliverables-oriented hierarchy that defines the work of the project and can be used on projects of any size or complexity.

4. B, C, D. The sections of a scope statement are project description, key deliverables, success and acceptance criteria, exclusions, time and cost estimates, assumptions, and constraints.

5. B. Decomposition breaks the major deliverables down into smaller, more manageable units of work that can be used estimate cost and time and perform resource planning.

6. A. The lowest level of a WBS is the work package. The number of levels will vary by project and complexity.

7. E. All of the options describe and provide examples of influences.

8. D. The scope management plan, not the WBS, describes how the deliverables will be validated.

9. D. The code of accounts identifier is a unique number assigned to each component of the WBS. It is documented in the WBS dictionary and is tied to the chart of accounts.

10. C. The first level of the WBS is the project name, in this case ABC Product Launch. The second level of the WBS represents major project deliverables, project phases, or subprojects. If the project has phases or subprojects, these are listed at the second level, with deliverables listed at the third level. Since the question asks about phase 2 of the project and option C is project phases, this is the correct second-level entry for the WBS.

11. **A.** The scope management plan contains a definition of how the deliverables will be validated, but the acceptance criteria are documented in the scope statement.

12. **B.** Scope creep involves changing the product or project scope without regard to impacts to the schedule, budget, and/or resources. KPIs are key performance indicators that help you incrementally monitor project performance.

13. **A, C, E.** They are budget, scope, and time, all of which impact quality.

14. **C.** Scope creep and change requests are examples of influences on the project. Influences can change, impact, or bring about new constraints.

15. **B.** Project planning processes are iterative, meaning you'll define the scope statement and other planning documents, and as you create these documents, more information may come to light or you may discover an element you missed. So, you'll go back through processes you've already started and modify them with the new information.

16. **A.** This is an example of acceptance criteria for the deliverable.

17. **B, D.** Work that is not included in the WBS is not part of the project. Exclusions from scope are work components that are not included in the project and should not appear on the WBS.

18. **C.** Whenever problems arise on a project that are outside the authority or control of the project manager to resolve or when problems have the ability to affect project outcomes, the sponsor should always be informed.

19. **A.** This is an example of a constraint because it dictates the actions of the project team.

20. **B.** The Planning process group is where you begin to define important documents such as the scope statement and project plan.

Chapter 5: Creating the Project Schedule

1. **B.** The critical path is the longest path on the project. The tasks have zero float because the critical path controls the project end date. Using critical path, you can determine which tasks can start late or go longer than planned without impacting the project end date.

2. **A.** A requirement such as weather conditions or a specific season that drives the scheduling of a task is an example of an external dependency.

3. **D.** Analogous estimating is also called top-down estimating. It is used early in the project, when there is not enough detail to do a detailed estimate.

4. **B.** There is a finish-to-start dependency relationship between Activity A and Activity B. You do not have enough information to determine whether the dependency between the two activities is mandatory, discretionary, or external or if they are critical path activities.

5. A, B. PERT charts and Gantt charts, a type of bar chart, are the most common ways to display a project schedule.

6. A. Duration compression involves either crashing the schedule by adding more resources or creating a fast track by working activities in parallel that would normally be done in sequence.

7. C. Float time is the length of time a task may be started late or the additional duration a task may take without impacting the project completion date. The early start and late start dates are the same, and the early finish and late start finish dates are the same.

8. B. A quality gate is used to determine whether the work is accurate and meets quality standards.

9. C. The total work day duration is eight days. The first day counts as one full day.

10. A. Float is always zero for the critical path activities, so early start and late start are the same date.

11. A. Fast tracking is a technique where you perform multiple tasks in parallel that were previously scheduled to start sequentially.

12. B. If you didn't know the quantity and rate, option C or D would be acceptable. In this case, you'd use the parametric estimating technique because you do know the quantity and rate; 30 hours times 4 miles is a total duration of 120 hours.

13. E. Governance gates might include go/no-go decisions, client sign-off, management approval, and legislative approval.

14. C. The task will take 15 business days to complete (not counting weekends or holidays) so that makes the task completion date the 22nd. Day 1 counts as the first full day of work.

15. D. Milestone charts list the major deliverables, key events, or project phases and show the scheduled and actual completion dates of each milestone. They may include other information, but that information would not be displayed as bar charts.

16. A. After the WBS is developed, the next step involves creating an activity list that describes the activities required to complete each work package on the WBS.

17. B. Once the schedule is approved by the sponsor, customer, and stakeholders, the schedule baseline is set. The baseline is the approved project schedule and shows task start and end dates and resource assignments.

18. A. Finish-to-start is the most commonly used logical relationship in network and schedule diagrams.

19. B. Eliminating Task B leaves you with the longest path through the network diagram, which is path A-C-E for a duration of 20 days.

20. C. Tasks A, B, D, and G represent the longest dependent path through the network diagram at 41 days.

Chapter 6: Resource Planning and Management

1.   A. When a new team member is introduced on the project, the team development stage starts again at the forming stage, no matter which stage the team was in before.

2.   A. Team-building activities help to build effective and efficient teams, improve morale, and build social bonds. Trust building will help form high-performing work teams but doesn't directly lead to building social bonds or effective and efficient teams.

3.   C, E. This situation describes varying work styles, which are a common cause of conflict. An interproject resource contention is where resources are working on multiple projects and there are scheduling issues. Low-quality resources lack skills or abilities. The negotiating conflict technique often uses a third party to help the two conflicting sides reach a resolution.

4.   B. The stages of team development are forming, storming, norming, performing, and adjourning. Confronting is a conflict-resolution technique.

5.   B. This question defines interproject resource contention. A resource shortage would occur if there was only one resource available or resources were scarce. Since this question describes more than one project, it involves an interproject resource contention.

6.   D. The smoothing conflict resolution technique is temporary, and one of its characteristics is emphasizing the areas of agreement and keeping the real issue buried.

7.   C. Negotiating is a technique that involves listening and asking questions.

8.   B, D, E. The conflict-resolutions techniques are smoothing, forcing, compromising, avoiding, and negotiating. Adjourning, norming, and storming are stages of team development.

9.   D. To address the issue, you need to understand what is behind the system engineer's current behavior. He may have been given additional work that you are not aware of, or he may misunderstand the project goals, to name just a couple of possibilities. The situation cannot be ignored, no matter how valuable the person is, and it should be handled in private.

10. A, C, E. A RACI is a matrix-based chart that shows resources (or business units) responsible for project tasks. It stands for responsible, accountable, consulted, and informed. Accountable does mean this resource approves the work, but the A in RACI stands for accountable.

11. D. Rewards and recognition do not have to involve money, and many times they may include rewards such as a thank-you, a letter to the functional manager, a public mention of the accomplishment at a team meeting, and so on.

12. A. This question refers to a resource shortage. You have one resource that's needed for two tasks. Resource allocation is assigning the resource with the right skills and abilities to the task. Interproject dependencies rely on one project finishing before the next can start, and shared resources are typically resources that are shared among departments or between the functional manager and the project manager.

13. B. The kickoff meeting is where the project team members and stakeholders are introduced to each other, and it's held at the beginning of the Executing process group.

14. C. The RACI chart acronym stands for responsible, accountable, consulted, and informed. The person accountable is also an approver of the work.

15. A, B. Whenever a new team member is introduced, the team development stage reverts to the forming stage and progresses through all the stages once again with the new team member.

16. B, C, E. Personality clashes and staff changes are situations where team-building activities can assist in solving problems. Organizational changes require immediate communication from the project manager. As a rule, most people are generally sensitive to change and are asking, "What does this mean for me?" This has a tendency to disrupt working patterns and decrease efficiencies, and it requires that you act as a change agent—getting people through the change while continuing the work of your project. Additionally, it's quite possible that an organizational change may directly affect your project, in which case you, too, need to ask, "What does this mean for the project?"

17. E. The WBS is too detailed to review at a project kickoff meeting and is better handled during a meeting with project team members only.

18. D. Smoothing is a lose-lose technique, forcing is win-lose, confronting is win-win, and avoiding is lose-lose.

19. B. Benched resources are costly to an organization. These are resources who are not currently assigned to project tasks and are typically between projects. This generally occurs in a projectized organization.

20. C. An organization breakdown structure shows work by the department or work unit responsible for completing the work packages. A resource breakdown structure shows the types of resources needed and the work packages. A project organization chart shows the hierarchy of the project team members, and an organizational chart shows the hierarchy of the reporting structure within an organization.

Chapter 7: Defining the Project Budget and Risk Plans

1. C, F. The labor is $75/hour times 10 trees is $750. This is the parametric method of estimating because you are multiplying the quantity times the rate. Analogous estimating involves using estimates from similar projects, and three-point estimates use the average of three estimates.

2. B. Top-down estimating is another name for analogous estimating.

3. B. A work effort estimate or person-hour estimate is used to develop the cost estimates. This is the amount of time it will take to complete the task from beginning to end without accounting for work breaks, holidays, and so on. Duration estimates account for holidays, work breaks, and so on. Bottom-up estimates are estimates for individual components of work that are rolled up into the overall estimate, and parametric estimates are usually derived by multiplying quantity by rate.

4. C. A contingency fund is an amount allocated to cover the cost of possible adverse events, and the project manager generally has the ability to use this fund. The project manager does not usually have the authority to spend money from the management reserve. The chart of accounts is a description of the accounts listed in the accounting ledger, and the cost baseline is the total expected cost for the project.

5. A. Management reserves are not part of the project cost budget or cost baseline.

6. C. Bottom-up estimates start at the work package level of the WBS. Each work package on the WBS for the first phase of the project is summed to come up with an overall estimate for this phase. Historical data would be useful if you were using the analogous estimating technique. The chart of accounts doesn't help at all with this exercise, and the scope statement will give you an understanding of what's detailed on the WBS, but it won't help with estimating.

7. B. Bottom-up estimates are the most accurate estimates, and analogous estimates are the least accurate. Estimates based on expert judgment are analogous estimates. Parametric estimates are only as accurate as the data you're using for the parametric model.

8. C. The cost baseline is approved by the project sponsor, not the project manager.

9. D, G, H. Three-point estimates are the average of the most likely, optimistic, and pessimistic estimates.

10. C. The burn rate is typically calculated using the cost performance index (CPI). This tells you the efficiency or benefits of the money spent at any point in the project.

11. B. The rate that is established for a given resource times the work effort (usually expressed in hours) will yield the total estimate for the task.

12. D. Transfer is a risk strategy that transfers the consequences of the risk to another party.

13. A. SWOT stands for strengths, weaknesses, opportunities, and threats. Examining the project from each of these perspectives helps you identify risks. The other options are cost performance measurements.

14. A, B, C. During the early stages of risk planning, a risk register typically has a risk identification number, a description of the risk, the probability and impact of the risk event, risk score, and risk owner. Your risk register could also contain the risk trigger and other pertinent information about the risk.

15. D. The risk strategy of accepting a risk involves dealing with the consequences when they occur. You don't prepare a risk response plan for risks you plan to accept.

16. B. The difference between what you planned to spend and what was actually spent is known as a budget variance.

17. A. Risk triggers are symptoms or signs that a risk event is about to occur.

18. C. Sharing is a positive risk strategy. The negative risk strategies are avoid, transfer, mitigate, and accept.

19. A. When project costs are displayed graphically over time, they represent an S curve. This is because spending starts out slowly on the project, picks up speed during the middle of the project, and tapers off at the end.

20. C, F. The risk score is calculated by multiplying the probability by the impact. Probability is the likelihood a risk event will occur. It is expressed as a number from 0.0 to 1.0. Impact is the consequence of the risk event if it occurs and can also be expressed as a number from 0.0 to 1.0.

Chapter 8: Communicating the Plan

1. C, E, F. A communication plan is developed to determine who needs communication, when, in what format, and the frequency of the communications. Once the plan is developed, it's used to update stakeholders, team members, vendors, and others who need information on the project.

2. D. In the sender-message-receiver model, the receiver is responsible for understanding the information correctly and making certain they've received all the information.

3. A. There are four participants in the meeting and six lines of communication. The formula for this is $4(4 - 1) \div 2 = 6$.

4. C. Gate reviews are a communication trigger. Language barriers, cultural differences, and others are factors that influence communication methods.

5. B. Frequency, level of report detail, types of communication, confidentially constraints, and tailor communication styles are all stakeholder communication requirements.

6. A. One of the purposes of a kickoff meeting is to introduce team members to each other. Video conferencing would be the best choice so that team members can see each other during introductions as well as hear the project goals and so on.

7. G. All of the options are considered communication methods. You should tailor the method of communication to the audience.

8. C. Communications planning is the process of identifying who needs to receive information on the project, what information they need, and how they will get that information.

9. B. The only factor influencing communications listed in the options provided is personal preference. All the incorrect options are communication methods.

10. D. Intraorganizational differences affect different departments across the organization. This question states you work for the PMO and the finance department is uncooperative, meaning you have two departments involved in the difference.

11. C. A stakeholder change is an example of a communication trigger.

12. B, E, F. The basic communication model is the sender-message-receiver model.

13. A. A face-to-face meeting with the employee is the best method of communicating when you need to discuss sensitive information. If the employee refuses to change their behavior, the next step might be meeting with the functional manager. However, you should always try to resolve the problem first with just the employee.

14. E. Technological factors are factors that influence communication methods.

15. D. Informal communications include email, hallway conversations, and phone calls. They are typically unplanned and casual in nature.

16. C. Virtual meetings can be attended by people from any geographic location. They are not in-person meetings.

17. B. The first step is to tailor the communication method based on the content of the message. There isn't enough information in the question to determine the content of the message, so we don't know which method is best to use.

18. A, F. Complex information is best delivered in a written format and then explained at an in-person meeting so the stakeholder can ask questions and you are able to determine whether their body language indicates they understand.

19. B, D. These are all stakeholder communication requirements and should be recorded in the communication plan.

20. A. The question describes factors that influence communication methods.

Chapter 9: Processing Change Requests and Procurement Documents

1. D. The customer or a stakeholder may have requested the new features. If these are required features that were omitted from the original scope statement, you need to analyze the impact to the project and obtain approval for the change. If you just make adjustments to the budget and schedule without any analysis, not only do you risk being late and over-budget, but there may be impacts to other areas of the plan or risks associated with this change. Removing the new features may add cost and time to the schedule as well as create a potentially hostile relationship with the customer. Unless this is a situation where the programmer has repeatedly changed scope outside of the approval process, requesting a replacement resource is not an appropriate response.

2. C. Corrective actions, defect repairs, and preventive actions are all types of change.

3. B. The change control board (CCB) is responsible for reviewing and approving, denying, or deferring change requests.

4. A. Emergency change request procedures should be documented so that changes that must be made on an emergency basis prior to the next CCB meeting can be made. All changes should be documented and reported at the next CCB meeting.

5. D. After options A–C are conducted, the change request and analysis are given to the CCB to make a decision. The appropriate project planning document is not updated until the CCB makes a decision regarding the disposition of the change request.

6. A, D. Determining the constraint that stakeholders think is driving the project will help you determine the kinds of trade-offs or alternatives you can propose to lessen the effect of the proposed scope change.

7. C. The correct action to take in this situation is to submit the change request to the CCB. If it is approved, it will require that you rebaseline the schedule to reflect the new dates.

8. D. Make-or-buy analysis is the technique of determining the cost-effectiveness of procuring goods or services outside the organization.

9. C. An acquisition gives power to the organization that is taking over. In this scenario, your old company has experienced some layoffs and managers from the new organization have taken over. This describes an acquisition.

10. D. The statement of work (SOW) describes in detail the goods or services you are purchasing from outside the organization.

11. B. Bidders conferences are usually set up shortly after the RFP is posted and allow vendors the opportunity to ask questions about the project.

12. A. Weighted scoring models weigh various criteria from the RFP and SOW, which allows you to score each vendor on each of the criteria and determine an overall score for each vendor.

13. D. Cost-reimbursable contracts are the riskiest for buyers, since the buyer is responsible for reimbursing the seller on the costs of producing the goods or services.

14. A. Time and materials contracts are a cross between fixed-price and cost-reimbursable contracts. They assign a unit rate for work, but the total cost isn't known until the work is complete.

15. B. The project management plan serves as the baseline for project progress and is used throughout the Executing and Monitoring and Controlling phases to determine whether the project is on track. It is used to help evaluate changes against the original goals and objectives of the project and serves as a communication tool.

16. A. The product owner is responsible for the backlog and communicating with the project stakeholders.

17. B, C. Standups and Scrum meetings are held daily to establish progress on the project. They are typically limited to 15 minutes in length.

18. C, D. Waterfall and Agile use iterative approaches to manage projects. PRINCE2 divides projects into stages, and the critical path method is a scheduling method.

19. C. Agile allows for continuous feedback and utilizes self-organized and self-directed teams.

20. B, E. Backlogs are used in the Agile methodology. The product owner prioritizes the backlog based on value to the business. Team members choose backlog items to work on in the sprint planning meeting. They may also break down large backlog items into smaller work components that can be completed in a single sprint.

Chapter 10: Project Tools and Documentation

1. C. The knowledge management tools include intranet sites, Internet sites, wiki pages, vendor knowledge bases, and collaboration tools. A dashboard is a reporting tool.

2. D. There is valuable information to be gained from a review of any project, even projects that do not complete. The assessment should focus on those phases of the project that did finish, as well as a look at whether anything could have been done differently to make the project a success. The purpose of lessons learned is not to assign blame, even for projects that are canceled.

3. B. Starvation is a project ending caused by resources being cut off from the project. Extinction occurs when the project work is completed and is accepted by the stakeholders. Addition occurs when projects evolve into ongoing operations, and integration occurs when resources are distributed to other areas of the organization.

4. B. Key performance parameters are used to measure performance or operational goals for systems. They are similar to KPIs, which are used to measure any element of the project or operational areas of the business to determine whether goals are being achieved. A balanced score card is a type of management tool used to determine whether organizational goals are being met.

5. A. Extinction occurs when the project work is completed and is accepted by the stakeholders. This is the best type of project ending. Starvation is a project ending caused by resources being cut off from the project. Addition occurs when projects evolve into ongoing operations, and integration occurs when resources are distributed to other areas of the organization.

6. D. A sign-off is the formal acceptance of the project's final product, service, or result. Its primary purpose is the customer's acceptance of the product of the project. Team members are released after sign-off, but this isn't the primary purpose of a formal sign-off. Both the project manager and the project team members may continue to be involved in the project until all closure activities are complete.

7. C. A Pareto diagram is a type of histogram that measures the frequency of occurrences of data elements in rank-order over time.

8. A. Both the successes and failures of a project need to be documented in the lessons learned report. Successes will provide blueprints to follow on future projects, and failures will alert teams on what to avoid. A good lessons learned document covers all aspects of the project from all participants. It should include all project information, not just schedule, budget, and changes, and it should never place blame for the things that went wrong.

9. D. Both team members and functional managers need to know in advance when you think a team member will be released. Team members may roll off the project at different times, so you need to discuss the release with each team member individually.

10. C. The fishbone diagram is a cause-and-effect diagram. Brainstorming sessions are a great way to construct this chart and determine what causes are impeding your results.

11. B. Action items generally arise during the status meetings. They should be documented in an action item list. They are assigned an identification number, a description, and an owner, and their status is recorded and reviewed at status meetings. Issues generally impact the project work directly and may impede progress or bring about a risk. Action items are usually "to dos" or questions that must be answered regarding the project.

12. A. Status reports update stakeholders on project progress. The issues log and action item list are usually distributed with the status report. Wiki pages are a way to distribute the reports. Meeting minutes document what occurred during the meeting and what decisions were made, while dashboards are up-to-date, real-time information on the status of key project elements.

13. D, E. Addition occurs when the project evolves into ongoing operations. Integration occurs when the resources on the project are reassigned to other projects or activities. When a project fails, is canceled, or otherwise ends before completion, the next step is performing a postmortem review.

14. B, D. Fishbone diagrams are cause-and-effect diagrams. Scatter diagrams are used to determine whether there is a correlation between cause and effect.

15. C. The project close report is produced at the end of the project, and it serves as the final status report. It summarizes the project goals, costs, schedule, lessons learned, and historical data.

16. B, C. If you have a sponsor who opts to cancel the project, you will still perform project closing procedures. During this process, you'll assemble the closure documents, perform a lessons learned analysis, and release any resources working on the project.

17. A. Executives don't have time to attend daily standup meetings and aren't likely to read daily emails. The best option is to provide them a dashboard with up-to-date information that's easy to read.

18. A. Team members can be released prior to the lessons learned session. If your team members are leaving the organization or are located at a different geographical location, you could perform a lessons learned session with them before they leave, or you could include them in the final lessons learned session using video conferencing or similar technology.

19. C. This question describes an issue that has occurred on the project that will likely impede progress. You will record this issue in the issues log and report regularly on its status.

20. D. The sponsor is the one who signs off on the closure documents. As the project manager, you create the documentation and provide supporting artifacts to demonstrate that all deliverables have been successfully completed.

Index

Note to the Reader: Throughout this index **boldfaced** page numbers indicate primary discussions of a topic. *Italicized* page numbers indicate illustrations.

D

Comprehensive Online Learning Environment

Register on Sybex.com to gain access to the comprehensive online interactive learning environment and test bank to help you study for your CompTIA Project+ certification.

The online test bank includes:

- **Assessment Test** to help you focus your study to specific objectives
- **Chapter Tests** to reinforce what you learned
- **Practice Exams** to test your knowledge of the material
- **Digital Flashcards** to reinforce your learning and provide last-minute test prep before the exam
- **Searchable Glossary** gives you instant access to the key terms you'll need to know for the exam

Go to http://www.wiley.com/go/sybextestprep to register and gain access to this comprehensive study tool package.

30% off On-Demand IT Video Training from ITProTV

ITProTV and Sybex have partnered to provide 30% off a premium annual or monthly membership. ITProTV provides a unique, custom learning environment for IT professionals and students alike, looking to validate their skills through vendor certifications. On-demand courses provide over 1,000 hours of video training with new courses being added every month, while labs and practice exams provide additional hands-on experience. For more information on this offer and to start your membership today, visit http://itpro.tv/sybex30/.